Challenges in Virtual Collaboration

Videoconferencing, Audioconferencing, and Computer-Mediated Communications

The research described in this report was conducted in the RAND National Defense Research Institute, a federally funded research and development center supported by the Office of the Secretary of Defense, the Joint Staff, the unified commands, and the defense agencies under Contract DASW01-01-C-0004.

Library of Congress Cataloging-in-Publication Data

Wainfan, Lynne
Challenges in virtual collaboration : videoconferencing, audioconferencing, and computer-mediated communications / Lynne Wainfan and Paul K. Davis.
p. cm.
"MG-273."
Includes bibliographical references.
ISBN 0-8330-3700-5 (pbk. : alk. paper)
1. Telematics. 2. Teleconferencing. I. Davis, Paul K., 1943– II. Title.

TK5105.6.W35 2005
302.2—dc22

2004025329

The RAND Corporation is a nonprofit research organization providing objective analysis and effective solutions that address the challenges facing the public and private sectors around the world. RAND's publications do not necessarily reflect the opinions of its research clients and sponsors.

Cover design by Barbara Angell Caslon

Published 2004 by the RAND Corporation
1776 Main Street, P.O. Box 2138, Santa Monica, CA 90407-2138
1200 South Hayes Street, Arlington, VA 22202-5050
201 North Craig Street, Suite 202, Pittsburgh, PA 15213-1516
RAND URL: http://www.rand.org/
To order RAND documents or to obtain additional information, contact
Distribution Services: Telephone: (310) 451-7002;
Fax: (310) 451-6915; Email: order@rand.org

Preface

This report was developed as part of a larger project on aids to high-level national-security decisionmaking. It discusses the effects of the medium of collaboration (face-to-face, videoconferencing, audio-conferencing, or computer-mediated conferencing) on group processes and outcomes. Questions or comments are welcome and should be addressed to the authors at the RAND Corporation's Santa Monica, CA, office:

Paul K. Davis, project leader
(pdavis@rand.org)

Lynne Wainfan, principal author
(Lynne_Wainfan@rand.org)

The research was performed in the Acquisition and Technology Policy Center of RAND's National Defense Research Institute (NDRI), a federally funded research and development center (FFRDC) serving the Office of the Secretary of Defense, the Joint Staff, the unified commands, and the defense agencies.

For more information on the Acquisition and Technology Policy Center, contact its director, Philip Anton (Philip_Anton@rand.org), at RAND's Santa Monica office. More information about RAND is available at www.rand.org.

Contents

Figures

Tables

Summary

Purpose

Virtual collaborations are collaborations in which the people working together are interdependent in their tasks, share responsibility for outcomes, are geographically dispersed, and rely on mediated, rather than face-to-face, communication to produce an outcome, such as a shared understanding, evaluation, strategy, recommendation, decision, action plan, or other product. This report summarizes the research literature on how the processes and outcomes of virtual collaborations are affected by the communication medium, discusses how problems in such collaboration can be mitigated, and suggests a strategy for choosing the most effective medium as a function of task and context. The focus is on interactive collaborations in real or near-real time.

Background

Virtual collaboration uses "mediated-communication" rather than face-to-face (FTF) encounters. The principal modes are audioconference (AC), videoconference (VC), and computer-mediated communication (CMC). These are increasingly being used for "reach-out" (e.g., contacting experts not readily available for FTF meetings), to improve response time, and to save money. Such forms of collaboration occur in many walks of life and at many different levels of organization. President Bush has critical meetings with cabinet members in which some participants attend by videoconference. Business-world chief executives do the same. During military conflicts and civil crises, staffs are commonly involved in virtual collaborations using, e.g., e-mail, Web-based chat rooms, other forms of CMC, AC, and sometimes VC. These may be deliberate (as, for example, in discussions a few days prior to a decision to go to war or in meetings held by a NASA mission director before launching or deorbiting spacecraft) or time-critical (as, for example, in war when new and fleeting targets arise and aircraft may be diverted from other tasks to strike them *if* prompt assessments indicate that doing so will be effective and will not result in unacceptable collateral damage; or when decisions must be made about where to send emergency response teams in a civil crisis with many simultaneous reports and a great deal of confusion). Clearly, it is important to understand how the form of virtual collaboration (i.e., the communication medium) influences group processes and outcomes—not only in obvious ways, such as timeliness, but with respect to issues such as the quality of outcomes.

Most research in mediated communication focuses on comparing a given medium with FTF communication or compares different media for specific tasks, such as negotiations. In this report, we step back and evaluate virtual collaboration for a broad range of task types and across VC, AC, and CMC. To our knowledge, it has been more than 25 years since the last paper (Williams, 1977) provided a comparably broad view.

This is an opportune time for such a review because of the proliferation of virtual collaboration and the related use of mediated communication. It is known that the various communication media can have significantly adverse effects, which are often not evident to participants. Can these adverse effects be avoided or mitigated by the appropriate choice of communication medium for the tasks at hand and by adopting good practices and aids? The media options here are improving because of advances in technology, including the ongoing convergence of VC, AC, and CMC in affordable systems. As a result, virtual collaborators can more often choose how to "meet," and—if they think to do so—how to "tune" effectiveness by astute relative emphasis on video, audio, and computer-mediated media in a hybrid session. We hope that this report will assist those who use virtual collaboration in choosing and tuning wisely.

Lessons on the Effects of Communication Media

How do the various communication media affect group processes and outcomes? More than 40 years of research have produced thousands of papers, books, and dissertations, but inferring an answer from that research is not straightforward, for many reasons. First, technology has advanced markedly, and older research results must therefore be evaluated critically for today's contexts. Second, there exists no agreed-upon model of virtual collaboration and mediated communication to help organize research findings. Third, outcomes depend on many factors: communication medium (or media), task type, context, group characteristics, and individual characteristics. Moreover, these factors interact, complicating interpretations of research. Finally, it is important to note that the baseline for comparison—FTF communications—also has many problems that have been studied in depth in recent decades. The resulting insights must be considered when organizing assessments of the mediated communications. For example, it is necessary to distinguish among meetings (whether FTF or virtual) that are relatively more focused on fact-finding, negotiations, socialization, or "people problems."

Despite these complications, a number of conclusions emerge from the empirical literature. The following have arisen consistently across different experimental conditions, and we see them as largely valid:

- All media change the context of the communication somewhat, generally reducing cues used to (1) regulate and understand conversation, (2) indicate participants' perspective, power, and status, and (3) move the group toward agreement.

In VC, AC, and CMC, participants tend to cooperate less with those at other "nodes" and more often shift their opinions toward extreme or risky options than they do in FTF collaboration:

- In VC and AC collaboration, local coalitions can form in which participants tend to agree more with those in the same room than with those on the other end of the line. There is also a tendency in AC to disagree with those on the other end of the communication link.
- CMC can reduce efficiency (as measured in time to solution), status effects, domination, participation, and consensus. It has been shown useful in broadening the range of inputs and ideas. However, CMC has also been shown to increase polarization, deindividuation, and disinhibition. That is, individuals may become more extreme in their thinking, less sensitive to interpersonal aspects of their messages, and more honest and candid.

Suggestions for Improving the Effectiveness of Virtual Collaboration

The empirical findings discussed above have some direct practical significance. Simple awareness of the tendency to form local coalitions, for example, might help prevent them. However, many of the effects discussed are not obvious to the participants, and explicit mitigation strategies are therefore called for. Drawing upon the literature and our own experience and reasoning, we summarize in Chapter Four a number of measures to mitigate problems. They include, for example, assuring that people know each other personally before relying upon virtual collaboration or, next best, building in time for "ice-breaking," socialization, and development of common understanding of purpose. Group leaders can be educated about how to prepare for, lead, or moderate virtual collaboration and about the problems to watch for and deal with. This typically requires conscious effort and experience, as does learning to be a good chairperson of a live meeting. Simply "doing it" and relying upon intuition is unwise.

Professional facilitators can be useful, although in our experience, they must have some subject-matter knowledge to have credibility with the team. In some cases, it is better to have a knowledgeable, respected, and "open" leader, trained in facilitation techniques appropriate to the communication medium. Using up-to-date technology (e.g., good sound, good video, and sharing of documents and graphics) is important. Moreover, specialized group software (e.g., group decision support software, GDSS) can be very helpful when used well—a function of education and experience. Individuals can be given guidelines and cautionaries to minimize flaming, reduce tendencies to think poorly of people at other nodes, and encourage awareness of unintended risky shifts and the like.

In addition to suggesting these discrete measures, the empirical record clearly suggests that a given type of mediated communication helps in some situations and hinders in others. We therefore suggest a strategy, summarized in Figure S.1, for choosing the collaboration medium (or tuning the different aspects of a hybrid system with a mix, e.g., of video, audio, and text). The left column distinguishes among different objectives for a given collaborative session; the logic tree in the middle suggests the preference order in which one might choose, for a given objective, the different types of medium. The right column summarizes important challenges associated with each. Interestingly, FTF sessions are not always the best, and relatively "low-tech" CMC characterized, e.g., by e-mail or chat rooms may be especially valuable in some cases. We believe that the strategy represented by Figure S.1 makes intuitive sense and accords with empirical work. However, the most important suggestion may simply be for virtual collaborators to consciously think about the kind of setup that would be most effective for their particular context, rather than simply following the path of least resistance or the path of usual procedure.

We end by again noting that concern about the matters we discuss in the report is clearly appropriate. Sometimes-subtle effects of the media used can indeed adversely affect high-consequence decisions made by virtual collaborators. These adverse effects include biases of judgment, the shifting of team choices toward risky or

Figure S.1
Strategy for Selecting the Best Medium for Virtual Collaboration

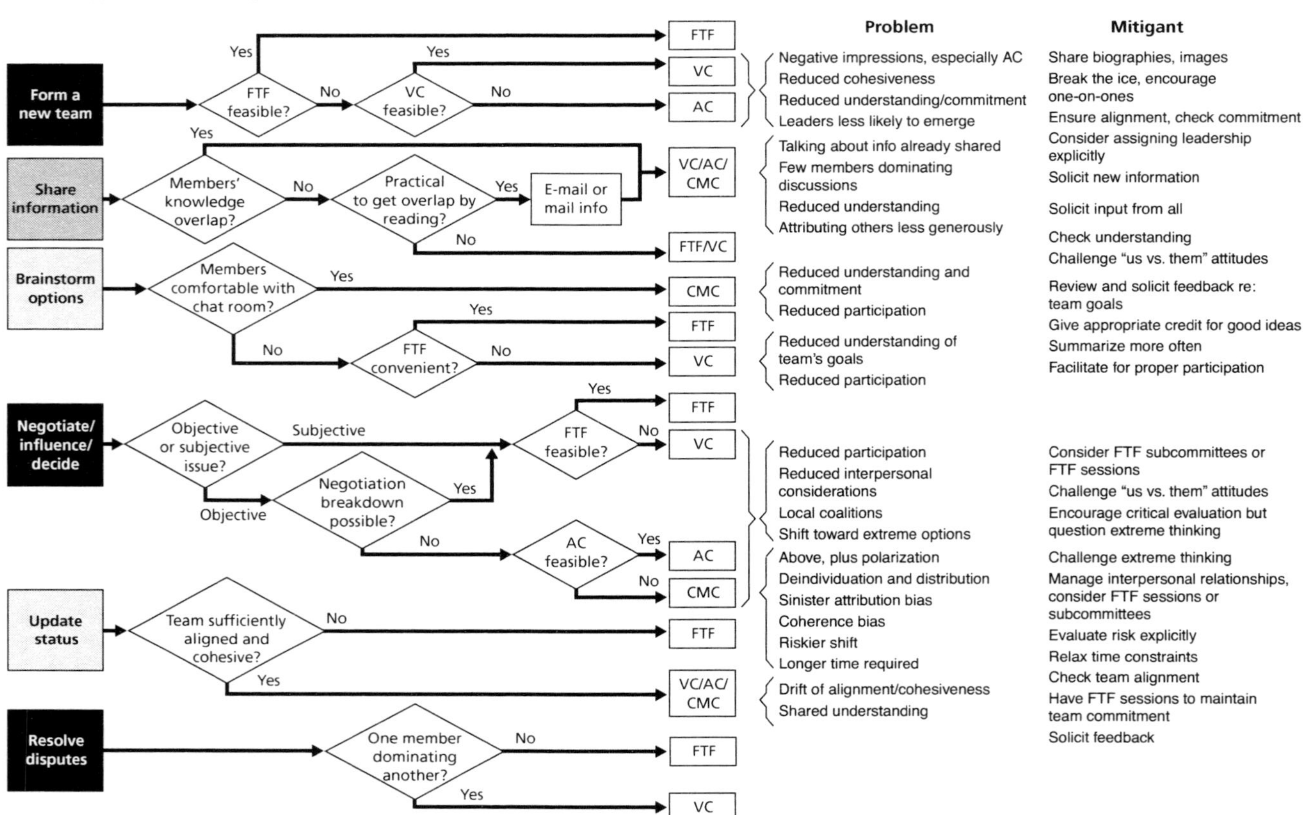

extreme options without adequate evaluation of risks, and the formation of negative attitudes about outside groups and participants who are more "distant" in the virtual collaboration.

Acknowledgments

The authors benefited from thoughtful technical reviews by RAND colleagues Tora Bikson and Ed Balkovich and from additional suggestions from Robert Anderson.

Acronyms

AC	audioconference
CMC	computer-mediated communication
GDSS	group decision support software
GSS	group support software
SIDE	social identity model of deindividuation effects
SMS	short-message service
VC	videoconference

CHAPTER ONE

Introduction

Background

An increasingly critical aspect of activity in modern workplaces is *virtual collaboration*, by which we mean (drawing upon a related definition by Gibson and Cohen, 2003) collaboration by people working together who are interdependent in their tasks, share responsibility for outcomes, are geographically dispersed, and rely on mediated, rather than face-to-face (FTF), communication.

Virtual collaboration occurs in a telecommunications network, each node of which contains one or more people. The nodes may be connected by any of several communication media, notably videoconferencing (VC), audioconferencing (AC), and computer-mediated communication (CMC).

Such virtual collaboration is important in a wide range of settings and at many levels of organization. To be sure, there are well-known reasons for preferring direct FTF conversation (President Bush has noted the importance to him of "looking them in the eye" when meeting with advisors before making high-consequence decisions, and business executives often think similarly).[1] Nonetheless, virtual meetings are increasingly common for any number of practical reasons, the most obvious of which are:

[1] See Woodward, 2002; Kelly and Halvorson, 2003; Berkowitz, 2003.

- *Broadening reach.* Virtual meetings can include participants who are geographically dispersed. These may be key aides, specialists with expert knowledge, respected advisors who could bring importantly different perspectives to the virtual table, or someone "on the scene" of developments under discussion.
- *Responsiveness.* Virtual meetings can be assembled far more quickly than physical meetings, and fast action can be essential in crises or competitions.
- *Adaptiveness.* New people can be added quickly, sometimes in a matter of minutes, when the need for their participation is recognized.
- *Time and money.* The direct and indirect costs associated with travel for FTF meetings can be substantial.

Unfortunately, the many benefits of virtual collaboration are accompanied by problems. Some of these stem from psychological effects associated with the communication medium, e.g., the emergence of animosities or in-group/out-group effects among participants, ineffective discussion, and the adoption of options that are riskier and perhaps less well-considered than those that would have emerged from FTF discussion.

Objectives of This Report

Against this background, the purpose of this report[2] is to review the empirical literature on how the communication medium affects virtual collaboration, to reason about the empirical observations, and to suggest ways to mitigate problems and exploit opportunities. An important feature of our review is that it covers all of the major media (VC, AC, and CMC) and a wide range of tasks. Thus, the review is

[2] A preliminary summary of interim results was presented and published as Lynne Wainfan and Paul K. Davis, "Errors Due to Virtual Collaboration," in Dawn A. Trevisani and Alex F. Sisti (eds.), *Proceedings of the SPIE*, Vol. 5423, *Conference on Enabling Technologies for Simulation Science VIII*, 2003. The current report, in addition to being much more detailed, has further material developed in 2004.

more comprehensive and synthetic than any previous efforts of which we are aware. The last comparably broad review appears to have been that done by Williams more than two decades ago (Williams, 1977).

Approach

We proceed as follows in this report. Chapter Two describes the methodology we used to search the literature on mediated communication. Chapter Three defines terms in some detail and then surveys the literature, drawing upon both empirical and theoretical work. Chapter Four is more speculative; it discusses ways to mitigate the various problems identified in the literature and suggests a strategy for using the range of collaborative methods wisely, as a function of what a particular group is attempting to accomplish. Chapter Five notes that much remains unknown and identifies fruitful paths for future research.

CHAPTER TWO

Definitions and Methodology

Videoconferencing, Audioconferencing, and Computer-Mediated Communication

Table 2.1 characterizes VC, AC, and CMC in simple terms. In VC, participants face a video image of another member or multiple images of other members. They may also use common graphics, such as a shared briefing or a shared whiteboard. In AC, participants are on the telephone with one or a number of other people. They may also use computer displays to see shared briefings or whiteboards. VC and AC may include subgroups meeting FTF in the same room. CMC is typically text-based, although it increasingly includes drawings, pho-

Table 2.1
Characterization of VC, AC, and CMC

Mode	Defining Characteristics	Examples
Videoconference (VC)	Useful real-time images and voices of other participants; may include other shared images/text.	Group videoconferencing in dedicated rooms; desktop videoconferencing.
Audioconference (AC)	Voice communication, but no useful real-time video images of other participants; may include other shared images, data, and text.	Phone calls, conference calls, or conference calls where people are also sharing views of images or documents.
Computer-mediated communication (CMC)	Text, images, and other data received via computer, without effective real-time voice or video images of other participants.	E-mail, chat rooms, discussion boards, text messaging, instant messaging, shared databases, application-specific groupware.

tos, and other images such as happy faces or "emoticons." CMC is either synchronous (i.e., chat rooms or instant messaging) or asynchronous (i.e., e-mail, discussion boards, application-specific groupware, or shared databases).

Relating the Types of Mediated Communications

Figure 2.1 shows an adaptation of Baltes et al.'s (2002) method of displaying the relationships among the various collaborative methods, showing each of the methods as points on a canvas defined by the

Figure 2.1
Placement of Communication Media, by Synchronization and Cues

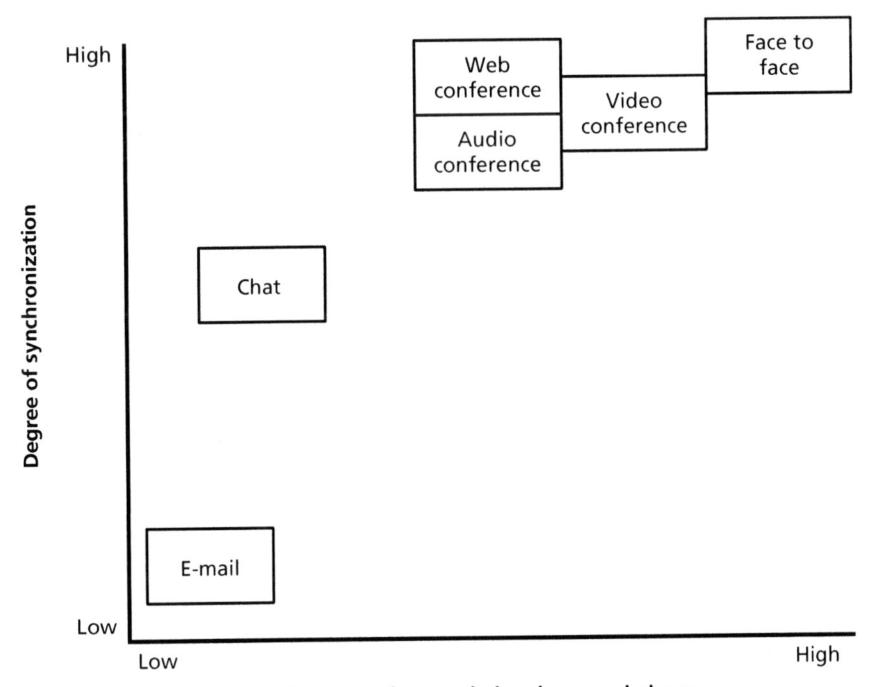

SOURCE: Adapted from Baltes et al., 2002.

presence of nonverbal and paraverbal cues (x axis) and the degree of synchronization (y axis). We have added a box for "Web conference," by which we mean the kind of meeting in which participants have a shared briefing or other document, and perhaps a shared whiteboard, that they discuss in real time with an audio link. This can be done with software such as NetMeeting,® LiveMeeting,® Timbuktu,® or WebEx.®[1]

Reconciling Findings Over a Period of Technological Change

An Initial Difficulty of Methodology

Our primary methodological challenge was that of pulling together findings from the literature, when the technology of virtual collaboration and mediated communication has been changing rapidly. We did not want to ignore early literature, but would results on early experiments of, say, VC still hold up, given the changes that have occurred?

To better appreciate this problem, consider the fact that research on mediated communication started almost 40 years ago, in the late 1960s, as AC became increasingly common in the workplace. VC came along somewhat later. Early AC technology allowed only one speaker at a time (half-duplex mode), and very-long-distance calls were sometimes subject to transmission delays. Similarly, early VC systems were jerky and updated the video frame slowly or had poor video and audio clarity. Early studies were "pure" in the definition of their domain (audio, video, or computer-mediated). Today, in contrast, all of the technologies are better, and various hybrid forms of communication are common, as discussed extensively in Chapter Three. How, then, do we pull together results from across the years? And does it make sense?

---

[1] NetMeeting and LiveMeeting, Timbuktu, and WebEx are registered trademarks of Microsoft, Netopia, and WebEx Communications, respectively.

To address the problem, we adopted a two-prong strategy: (1) establish a convention for defining the effective communication medium of a hybrid system, and (2) evaluate research results case by case.

Defining the Medium by the Form of Interpersonal Feedback

First, we decided to characterize the medium of communication in a given case by the dominant type of feedback participants receive in interacting *with each other*. That is, recognizing that the kinds of effects that turn out to be important in this report revolve around the nature of human interaction, rather than, say, the magnitude of analytical data in a session, we found that we could look at a given virtual collaboration experiment, which might involve a mix of audio, video, and computer-mediated communication, and characterize it overall as one of VC, AC, *or* CMC. This was a great simplification, but one that proved consistent with the research findings. That is, we make distinctions in terms of VC vs. AC vs. CMC vs. FTF, but the principles can then be applied reasonably well to hybrid systems as well, once one recognizes what kind of person-to-person feedback dominates a particular hybrid system. Figure 2.2 suggests this graphically, showing VC, AC, and CMC as archetypes. On the left, a *multimedia Web conference* might be a virtual collaboration with several teams meeting in rooms fitted with VC monitors and also equipped to project viewgraphs from a networked program, provide excellent audio, and perhaps more. A personal version of such a conference might involve dispersed small teams working at desktop computers fitted with miniature cameras. Participants would see on a given computer a window with a shared document, such as a briefing, and small, but high-quality, video images of other small teams. They would have excellent audio, either through a telephone link or as part of the audiovisual system. A Web conference (sometimes called a data conference) would be similar but would lack the video. Teams might be in dedicated rooms or at personal computers. Chat-room discussions might involve numerous individuals, rather than sizable teams, with the individuals interacting primarily through chat-room-style text, but with some attachments, such as photos or tables of data.

Figure 2.2
Mapping of Modern Hybrid Mediated-Communication Types into the "Archetypes" of VC, AC, and CMC

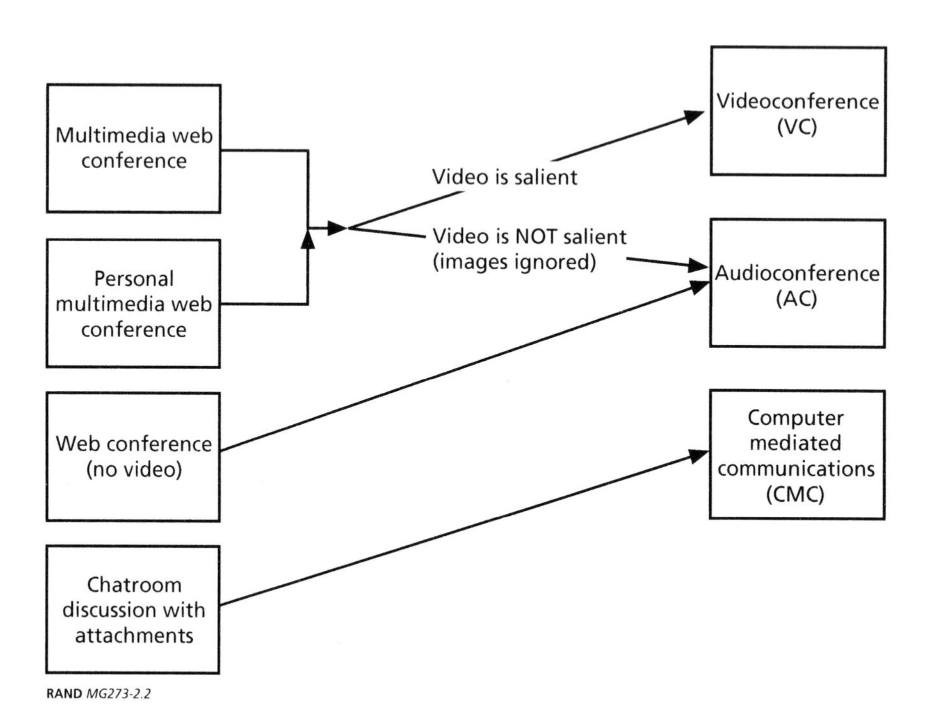

RAND *MG273-2.2*

Many other hybrid variations are possible, but these are representative. Moreover, we have personal experience with the first three and consider them important aspects of modern technology.

Case-by-Case Assessment of Older Research

The second prong of our strategy was to look critically at older research reports to assess as best we could how relevant they are for the modern world. This might have been quite troublesome and subjective, but in fact we found that many of the older papers remain unusually thoughtful and useful today, despite the changes in technology and practice. Moreover, we were surprised to find that the technology in some of the old studies is not as different from today's "normal" level as we had expected: Today's *typical* desktop VC quality is similar to older videoconference-room quality (although today's videoconference-room quality is much better), AC delays are still present in some mobile communications, and some data-intensive CMC remains similar (e.g., complex tables). Nonetheless, this case-by-case assessment did cause us to discard some of the older studies from our survey.

Methodology for Using Our Results

A spinoff of our conclusions about how to reconcile research results over a period of rapid technological change was a sharpened recognition that our results would be useful not only for assessing pure VC, AC, and CMC, but for "tuning" those components in modern hybrid forms of collaboration. That is, the vector of technological change is one of ongoing convergence of VC, AC, and CMC capabilities into desktop, and even mobile, computers. Indeed, in the future, it will be increasingly unusual to have pure VC, AC, or CMC. However, a given hybrid setup, e.g., one with all types of media, can be much more or less effective for a given purpose, depending on the relative emphasis of the various modes in the setup: Is video mere window dressing, or is it useful? Should it be emphasized or deemphasized? Is the meat of the virtual meeting in the focus on shared data and audio discussion? Or, when all is said and done, would something like an organized e-mail exchange or chat-room session suffice? We hope that our review of issues across modes will prove useful in making choices about collaboration mode or about how to tune hybrid modes (e.g., when to play up or deemphasize video). The choices will be task- and situation-dependent. Chapter Four presents our suggestions on these matters.

Having established these basic constructs, the next methodological issue was that of searching the literature and organizing results.

Search and Synthesis Methods

Defining Our Scope: Interdisciplinary, but Limited

Our review was broad not only in its time span, but also in the sense of touching upon a number of disciplines, notably psychology, communication and decision sciences, organizational behavior, business practices, and computer science. To narrow our scope, we focused on the tasks that support interactive collaboration between virtual team members in real or near-real time. For our literature review, we started with decisionmaking tasks in which groups communicate via technology. We expanded our aperture somewhat to include outputs other than decisions and to include the other processes that lead to a group's outcome: developing a shared understanding of the objectives; gathering information; identifying, evaluating, and selecting options; and negotiating consensus. We excluded some topics from our review, such as specific hardware and software evaluation, development of shared databases and other repositories, and tasks that involve a single person interacting with information (e.g., search methodology). We only lightly scanned the literature on mediated communication for personal or social tasks (such as social-issue discussion boards and chat rooms, multiplayer games, and electronic dating) and for broadcast-type tasks (distance learning, WebCast, broadcast speeches, newsgroups, etc.).

Search Methods Used

In searching the literature, we used a number of different queries or query methods, summarized in Table 2.2. The methods ranged from database searches to reference-list tree expansions to personal e-mail inquiries. Because it was not productive to search every permutation of the 21 keywords in all 13 databases addressed, we developed a "best-string" search methodology, where fruitful searches of keywords with Boolean operators were repeated in multiple databases. An example of a best search string is [(bias* OR error* OR problem* OR challenge* OR difficult*) AND (("face to face" AND group) OR ("computer mediated communication") OR (teleconference* OR "telephone conference" OR audio*) OR (video* NOT videotape))].

During our literature review, we often found it helpful to contact authors via e-mail for questions, clarifications, and advice. Table 2.2 also lists the researchers who responded to our inquiries. We found several reviews and meta-analyses of research in a particular mode, and these so-called "best papers" provided useful reference lists

Table 2.2
Summary of Literature Search Methodology

Databases searched	PsychInfo, ECO, Web of Science, Dissertations, Articles First, Papers First, WorldCat, ABI/Inform, ATSA, Science Direct, Computer Science Index, Communication and Mass Media Complete, Google
Keywords used	Communication media, computer-mediated communication, video-mediated communication, teleconferencing, audioconferencing, decisionmaking, cognitive biases, virtual teams, electronic groups, opinion shift, choice shift, polarization, disinhibition, virtual collaboration, virtual teams, mediated groups, face to face, communication mode, remote collaboration, electronic meetings, distributed work groups
Reference-list tree expansions	Using reference lists from so-called "best papers" (meta-analyses, reviews[a])
Follow-on works	Using the web of science database to find papers referencing "best papers"
Experts who responded to our e-mail containing follow-up questions	Takashi Tsuzuki, Ellen Isaacs, John Tang, Charles Liu, Byron Reeves, Clint Heinze, Brian Whitworth, Catherine Cramton, Ned Kock

[a]See Hollingshead and McGrath, 1995; Baltes et al., 2002; McLeod, 1992; Pinsonneault and Kraemer, 1989; Hedlund, Ilgen, and Hollenbeck, 1998; Hightower and Sayeed, 1995; Straus and McGrath, 1994; Vroman and Kovacich, 2002; Williams, 1977; Fjermestad, 2004; and Bordia, 1997.

and a high-level picture of how other researchers have structured their findings.

Organizing the Results

To organize our hundreds of results, we chose Pinsonneault and Kraemer's well-known 1989 framework for mediated communications, as discussed in Chapter Three.

Once the media effects were captured and organized within the framework, we began the effort of detecting effects common to all media. Although most previous reviews and meta-analyses focus on one mode, we were able to look across research on all the standard communication modes, comparing theoretical explanations and finding broad themes. This perspective helped us find factors to explain what appeared to be contradictory research findings but were instead interaction effects.

Synthesizing Strategies

The research consistently shows that the media used help some tasks and hinder others. Using the research literature, our own experience, and suggestions from our colleagues at RAND, we compiled a set of strategies to leverage the media's benefits and mitigate their adverse effects. Through a series of internal and customer reviews, we discussed the research findings and checked the strategies we had developed.

CHAPTER THREE

Virtual Versus Face-to-Face Collaboration: A Survey of the Literature

A Framework for Reporting Effects

Having defined types of virtual collaboration in terms of different forms of mediated communication, we now discuss how we will describe effects reported in the literature. We are interested in how the nature of collaboration affects both group processes and outcomes. What happens to group dynamics when one or more collaborator is in a different location or shares an affiliation such as country, company, or university? Is the quality of discussion higher or lower, and by what measure? Does communication mode affect people's influence on one another? What other effects might occur, e.g., the likelihood of consensus or opinion change? How do the media affect the group's output? Are biases and errors reduced or exacerbated?

To organize the research findings, we started with Pinsonneault and Kraemer's 1989 framework, for several reasons: First, this framework is typical of the context/process/outcome models used by many key researchers in the area (Hackman and Morris, 1975; Dennis et al., 1988; Finholt and Sproull, 1990; McGrath and Hollingshead, 1991; Fjermestad et al., 1993; Benbasat and Lim, 1993). Second, it is derived from and used by researchers performing meta-analyses of the appropriate research literature. Finally, the model's decision-making focus works well for virtual collaborations, which often produce an output such as a decision or a recommendation. However, Pinsonneault and Kraemer's model required major adjustment for our purposes. For one thing, it did not include factors that later were found to be affected by the communication medium (e.g., leadership emergence, polarization, disinhibition, local coalitions).[1] Also, their context/process/outcome framework was challenging due to the closed-loop nature of some collaborations; that is, the outcome state of one collaboration may be the context for the group's next collaboration. We considered non-context/process/outcome models (e.g., Slevin et al., 1998; Whitworth, 1998) but found that by adapting Pinsonneault and Kraemer's approach, we could more faithfully represent the majority of the research findings. Figure 3.1 shows the framework we used.

The first category, contextual variables, includes factors in the immediate environment of the group rather than in the broader organizational environment. The second category, group process, includes characteristics of the group's interaction. The last category, outcomes, includes task-related and group-related results of the collaboration.

Contextual Differences Between Face-to-Face and Mediated Collaboration

Several contextual differences between FTF and mediated communication have been shown to affect process and outcome. Group members' experience, attitudes, and comfort with the media (for instance, typing speed for text-based CMC) can interfere with or distract members from the task at hand. Alternatively, experienced users may focus more on the task than on social considerations, since extraneous cues are reduced.

The group structure in mediated communication is often broader yet more agile than that of an FTF team. Breadth comes from mediated communications' wider reach for subject-matter experts who can participate regardless of their location or time zone,

---

[1] We added the following factors to Pinsonneault and Kraemer's framework: group experience, breadth, agility, geographic distribution, and existing social networks; task consequences and time sensitivity; leadership emergence, trust, influence and persuasion, and disinhibition; biases and errors, choice or opinion shift, shift toward risky or extreme options; polarization; attitude toward other group members, including positive ratings, rapport, ability to perceive deception; cohesiveness, cooperation, and local coalitions.

Figure 3.1
Framework for Factors Affecting the Work Flow in Groups

Contextual Variables

1. Personnel factors
- Attitude
- Abilities
- Individual motives
- Background/experience

2. Situational factors
- Reasons for group membership
- Group development stage

3. Group structure
- Work group norms
- Status relationships
- Breadth, agility
- Location distribution
- Group size
- Anonymity
- Existing social networks

4. Technological support
- Degree and type
- Facilitator(s)

5. Task characteristics
- Complexity/uncertainty
- Nature/type
- Time sensitivity
- Consequences

Process Variables

1. Output characteristics
- Depth of analysis
- Participation
- Consensus-reaching
- Time to reach decision

2. Communication characteristics
- Clarification efforts
- Communication efficiency
- Information exchange
- Nonverbal communication
- Task-oriented communication

3. Interpersonal characteristics
- Leadership emergence
- Trust
- Influence and persuasion
- Domination by a few members
- Disinhibition

4. Structure imposed by media

Group Outcomes

1. Output characteristics
- Quality/variability w/time
- Biases and errors
- Choice or opinion shift
- Shift towards risky/extreme options
- Cost/ease of implementation

2. Attitude of group members toward the decision
- Polarization
- Comprehension
- Satisfaction/acceptance
- Commitment
- Confidence

3. Satisfaction with the group process

4. Attitude toward other group members
- Positive ratings
- Rapport
- Ability to perceive deception
- Cohesiveness
- Cooperation
- Local coalitions

RAND *MG273-3.1*

SOURCE: Adapted from Pinsonneault and Kraemer, 1989.

sometimes participating only as needed (Slevin et al., 1998). Agility comes from quicker response time and participants' ability to access additional information (or experts) more quickly and easily if they are participating closer to their work site. Eveland and Bikson's 1992 field study showed that CMC changes the structure of subcommittees, making them more elaborate and evolved over time and utilizing people on multiple subcommittees more than is done in FTF groups.

The media—including VC—filter cues such as facial expressions, gestures, vocal intonation, and indicators of understanding, as well as common ground, power, and status. Mehrabian (1971) reports that up to 93 percent of meaning is contained in facial and vocal cues rather than in text. Although that may be unusual in this era of briefings, it is surely true that miscommunication may happen more often, especially in CMC, when cues that indicate sarcasm or humor are attenuated. Another problem, especially with CMC, is the inherent ambiguity of English, for which we compensate in FTF collaboration with intonations, checks ("Do you know what I mean?"), exchange of anecdotes firming up or readjusting the points being made, facial expressions, and gesticulations (e.g., two fingers forming quotation marks). "Antagonyms" are particularly mischievous, since they have opposite meanings, depending on context. Examples include *anxious* (excited or filled with anxiety), *enjoin*, *general*, and *presently*.

The media also do not capture casual conversations in hallways and elevators, at social events, and in unplanned workplace interactions.[2] Finally, as we will see later, there is some evidence that media affect work-group norms—standards of group members concerning appropriate or inappropriate behaviors.

Power and status are harder to detect in mediated communication, because context cues such as seating position, office location, clothing, posture, and eye contact are reduced (Dubrovsky, Kiesler, and Sethna, 1991; Short, Williams, and Christie, 1976; Sproull and Kiesler, 1986). As we will see later, status effects can be exaggerated or mitigated, changing the content and distribution of discussion. Anonymity—possible only in CMC—can significantly affect group process and outcomes, for several reasons:[3] It reduces apprehension about performance, pressure to conform,[4] and inhibition (sometimes resulting in extreme comments, overly frank discussion, or intimacies[5]); it reduces the amount of influence an expert or dominant participant might otherwise have; and it diffuses individual responsibility, which can lead to group decisions that are riskier than the

---

[2] Kraut et al., 1993; Sproull, 1983; Whittaker, Frohlich, and Daly-Jones, 1994.

[3] See Vroman and Kovacich, 2002.

[4] See Weisband, 1992.

[5] See Kiesler and Sproull, 1992.

individuals' decisions. In addition, anonymous team members, feeling less visible, may exert less effort, resulting in reduced commitment to the team, productivity, and collaborative decisionmaking (Vroman and Kovacich, 2002). On the positive side, as we will discuss later, the use of anonymous CMC has been shown to produce more novel suggestions in group brainstorming,[6] increase participation by those who might otherwise remain silent during meetings, and allow users to review comments more objectively (Bikson, 1996).

Task characteristics affect group processes differently in mediated communication in several ways. First, a complex task may be made even more complex by the inability to adequately share flexible visual aids or to do something more ad hoc, such as moving to a large whiteboard.[7] Also, cognitive workload has been shown to be higher in VC and synchronous CMC than in FTF.[8] This can cause group members to shift to simpler problem-solving strategies that are not consistent with their training, be unable to raise counterarguments, or be more biased in their judgments.[9] Finally, communication media have been shown to affect processes and outcomes differently, depending on the type of task being performed. Most group tasks can be mapped into the four categories suggested by McGrath (1984, p. 61): (1) *generate*, (2) *choose*, (3) *negotiate*, and (4) *execute*, as shown in Figure 3.2.[10] Tasks in the *generate* category include agenda- and goal-setting, brainstorming, and devising potential solutions. The *choose* category includes tasks in which the group selects the "right" or "best" option among alternatives. *Negotiate* tasks allow the group to resolve conflicts. And *execute* includes factors such as time constraints

[6] See Fjermestad, 2004; Siau, 1995; Valacich, Dennis, and Connolly, 1994; Dennis and Valacich, 1993; Gallupe, Bastianutti, and Cooper, 1991.

[7] Technology is improving this situation with software such as NetMeeting and Timbuktu and with increased availability of broadband connections.

[8] See Hinds, 1999; Shamo and Meador, 1969; Graetz, 1997.

[9] See Hinds, 1999; Morley and Stephenson, 1969.

[10] In this context, intellective tasks, listed under *choose* in Figure 3.2, are those that require use of the intellect rather than the senses alone. This is the third definition given at www.webster-dictionary.org.

and decision quality that allow the group to produce a product. As shown in Figure 3.2, these tasks can be related to one another within a two-dimensional space where the horizontal axis represents the degree to which the task requires cognitive vs. behavioral activities. The vertical axis represents the amount of interdependence among group members, or the degree to which the interests of the group members are mainly in accord with those of other parties in the situation.[11] For instance, the *negotiate* task would lie on the "conflict-resolution" end

Figure 3.2
Task Types

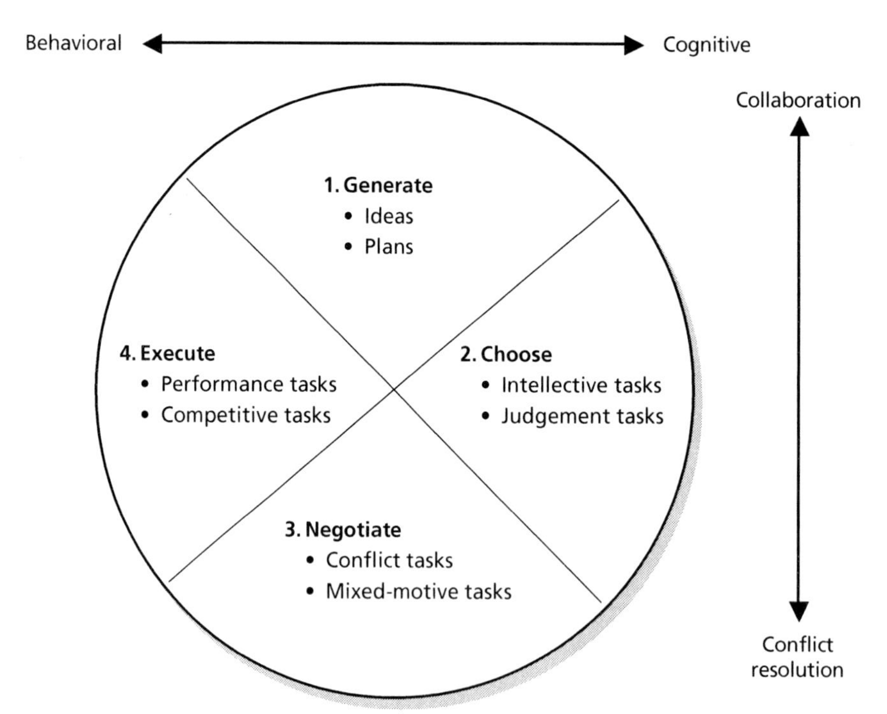

Adapted from McGrath, 1994.
RAND *MG273-3.2*

---

[11] Personal correspondence with J. E. McGrath, 2004.

of this axis. Some authors prefer to distinguish between *convergent* and *divergent* thinking. Convergent thinking involves bringing material from a variety of sources to bear on a problem in order to produce a "correct" answer. Divergent thinking, in contrast, involves creative elaboration of ideas. McGrath's *generate* category would involve divergent thinking, whereas *choose* would require convergent thinking. As we will see later, mediated communication helps some tasks and hinders others.

Process Differences Between Face-to-Face and Virtual Collaborations

Videoconferencing

Use of Videoconference Rooms. Consider first the "classic" videoconference, in which groups gather in dedicated conference rooms equipped with video equipment. We shall discuss desktop VC later.

Interaction patterns during classic VC are more similar to those of FTF communication than are such patterns in other types of CMC, but there are differences. Although VC image quality has improved since the early VC studies, it is still difficult to maintain eye contact due to image resolution and the distance between the camera and the monitor, and it is challenging to interpret body language and gestures, especially as the number of participants increases. Mediated communication—even VC—limits nonverbal, paraverbal, and status cues and reduces the "richness" of the information communicated. Studies show that VC participants may have difficulty identifying a remote speaker, detecting movements, attaining mutual gaze, and gaining floor control.[12]

Researchers have found that the discussion in VC (as well as in AC and CMC) tends to be less social and more task-oriented than in FTF.[13] Krauss and Bricker (1967) observed that 5 percent of

---

[12] See Daft and Lengel, 1986.

[13] See Krauss and Bricker, 1967; Isaacs and Tang, 1994; McLeod, 1992.

communication in VC meetings was social, non–task-oriented conversation, compared with 20 percent in FTF meetings. Consistent with this finding, other researchers have found VC discussion to be more orderly, formal, and polite[14] than FTF discussion, with a reduction of back-channeling and interruptions[15] and less conflict.[16] Nonetheless, according to Purdy and Nye (2000), VC is actually less time-efficient than FTF collaboration, possibly due to participants' difficulty in understanding each other and regulating conversations.[17] Participants also need more advance preparation to accomplish objectives in VC.[18]

There is evidence of less participation in VC than in FTF communication, perhaps due to dissatisfaction with current VC technology[19] or the "staged" feel of a videoconference-room meeting. Although there may be tasks in which participation is not critical to the outcome (such as information broadcasts), a virtual collaboration with its shared responsibility and interdependency would seem to be improved by more participation.

In groups with no established hierarchy, leaders emerge less often and participation and influence are more equal in VC groups than in FTF groups. In Strickland et al.'s 1978 study, VC groups showed weaker dominance hierarchies, and the hierarchies that did emerge were less stable over time.

Another effect of VC on group dynamics is a reduction in cohesiveness, or members' attraction to the group. Cohesive groups tend to have warm, personal, sociable interaction among members.[20] VC

---

[14] See Finn, Sellen, and Wilbur, 1997.

[15] See Sellen, 1995; O'Connaill, Whittaker, and Wilbur, 1993.

[16] See Barefoot and Strickland, 1982.

[17] See Straus, Miles, and Levesque, 2001.

[18] See O'Connaill, Whittaker, and Wilbur, 1993; Straus, Miles, and Levesque, 2001.

[19] See Yoo and Alavi, 2001; Anderson, Newlands, and Mullin, 1996; Chapman, 2001; Straus, Miles, and Levesque, 2001.

[20] See Yoo and Alavi, 2001.

participants have more intragroup conflict than do FTF groups,[21] perhaps because VC participants exchange fewer social remarks, are more dissatisfied with the medium, and participate less than they do in FTF communication. In our own experience, even when friends are participating and exchanging pleasantries, the "cybergulf" can be distinctly cooling.

One explanation for the reduced cohesiveness seen in VC may come from experiments on the media's effect on trust. Trust is a broad concept, defined by Mayer, Davis, and Schoorman (1995) as "willingness to be vulnerable, based on expectations about the actions of others" (p. 712). Bos et al. (2002) assigned teams to either FTF, VC, AC, or CMC conditions and asked them to solve a social-dilemma problem. This type of problem is commonly used to measure trust, since social dilemmas are defined as problems in which the best interests of the group conflict with the best interests of each individual (e.g., the prisoner's dilemma[22]). In Bos's experiment, the VC and AC groups did as well as the FTF groups in group payoff (the measure of trust), but the pattern of trust development was different. VC and AC groups took longer to reach cooperation (an effect Bos calls *delayed trust*) and exhibited *fragile trust*, i.e., repeated cycles where one player violates an agreement and others retaliate, followed by group discussion and reestablishment of trust.

At least one study has found that persuasiveness is lower in VC than in FTF. Ferran-Urdaneta (2001) attributes this to increased cognitive workload and decreased interaction in VC, which results in participants being less willing or able to engage in systematic elaboration of the message.

Small delays in VC's audio can frustrate participants and seriously disrupt their ability to reach mutual understanding.[23] Participants in Tang and Isaac's 1993 study found that the audio delay made it difficult to manage turn-taking and to coordinate eye con-

---

[21] See Barefoot and Strickland, 1982.

[22] The prisoner's dilemma is a classic problem of game theory discussed in standard texts.

[23] See Finn, Sellen, and Wilbur, 1997.

tact. They ended up turning off the audio and choosing a phone call (using speakerphones), even though the audio arrived before the video, the quality was poorer, and the speakerphone allowed only one party's sound to be transmitted at a time. This observation supports other research findings—and our own experience—that for most tasks, audio quality and responsiveness are more important for participant satisfaction than is video quality.[24]

Clearly, technology is improving, and classic VC will improve as well. High-ranking general officers have been using VC effectively for some years in settings as recent as the Gulf War.[25] They, however, have top-of-the-line equipment, such as large, high-resolution screens. We are unaware of any systematic effort to contrast this "richer" experience with the more usual VC experiences discussed above.

Personal VC. For a decade or so, it has been possible for two people seated at their personal computers to have a videoconference. In earlier years, however, this mode of operation did not "take off," in part because it just didn't pay its way: pictures were small, resolution was poor, and so on. A new wave of technology is now entering the market, however, and the situation may be changing. We have personally had good small-group meetings with two or three people huddled around desktop computers in different locations. It is now possible to view and operate on common screens (e.g., a PowerPoint presentation or a computer program), using software such as NetMeeting or Timbuktu, so the possibilities are changing. Our experience, however, has been in small project settings, and we did not find much research data of a more general kind. Newlands et al. (2000) found that collaborators using modern desktop VC elicited more listener feedback than did FTF participants, offered more information about their task and activities, and asked more yes-no questions. Newlands et al. explain that VC participants adapt their discussion to accommodate for the reduced visual feedback in VC

---

[24] See Gale, 1990; Ochsman and Chapanis, 1974; Adrienssen and van der Velden, 1993.

[25] See Woodward, 2002.

and spend more time explicitly checking for alignment and common ground. It remains uncertain that improved personal VC will be good for *creating* social ties. It does, however, seem to reduce the "cyber-gulf" relative to classic VC, and it can be quite effective for collaboration among those already acquainted and "teamed."

Audioconferencing

Audioconferencing removes all visual cues about other participants, reducing the ability to show understanding or agreements, forecast responses, enhance verbal descriptions, manage extended pauses, express attitudes through posture and facial expression, and provide nonverbal information.[26]

Several studies compare process differences between AC and FTF. Harmon (1995) reported typical challenges to adapting to audio-only communication: People look at, and gesture at, the speakers and have trouble with turn-taking, speaker identification, and interpretation of the discussion.

Burgoon et al. (2003) and Krauss (1976) found that lying may be easier to detect in AC than in FTF communication, possibly because there are fewer visual "distractions." Findings from Horn's 2001 VC study support this distraction theory; lie detection was significantly better when video quality was degraded. Two other studies, however, failed to replicate the finding that AC produces better lie detection (see Reid, 1970), although one found (as other studies have) that participants report lower confidence in AC judgment than in FTF,[27] and they believe that others agree and understand less in AC than in FTF.

Status seems to affect the AC process differently than it does in FTF communications.[28] In new groups forming via AC, the emergence of leaders is suppressed,[29] as is the case in VC. For established

---

[26] See Isaacs and Tang, 1994.

[27] See Young, 1974a, b.

[28] See DeSanctis and Gallupe, 1987.

[29] See Rice, 1984; France, Anderson, and Gardner, 2001.

groups, results are mixed. Harmon's and other studies show that a leader's dominance may be relatively impervious to mediated communication. However, France, Anderson, and Gardner's 2001 field study of an existing group of workers with an established hierarchy showed that AC technology actually *exaggerates* the impact of status. FTF meetings typically have unequal participation, with leaders talking more than individuals even when the leaders are not the experts on the subject.[30] By measuring the extent to which attendees in FTF and AC meetings initiated and engaged in conversations, France and his colleagues found that the most dominant individual in audioconferences took part in 45 times more pairwise conversations than did the least dominant individual. This dominance was approximately three times greater than in FTF meetings.

France, Anderson, and Gardner's (2001) theory of why low-status individuals find it difficult to contribute to AC discussions involves "the lack of non-verbal cues that can aid turn-taking, combined with (1) the participants' knowledge of the group's status hierarchy and (2) the tendency to compare oneself unfavourably to those of higher status" (p. 859). If this theory is correct, it has both positive and negative aspects. A dominant individual serving as a leader may help the group focus and reach consensus.[31] Equality of participation may be inefficient if knowledge or good ideas are unevenly distributed among group members, decision time is limited, or acceptance of a decision by group members is not critical.[32] However, AC's reduced interaction could be problematic for group understanding and consensus, for coordination of the group's goals,[33] or for certain types of tasks, specifically, problem-solving and innovation.[34]

---

[30] See Bales et al., 1951; Berger et al., 1977; Bales, 1950.

[31] See Hiltz, Johnson, and Turoff, 1986.

[32] See Vroom and Yetton, 1973.

[33] See Clark and Wilkes-Gibb, 1986.

[34] See Carletta, Garrod, and Fraser-Krauss, 1998.

Computer-Mediated Communication

In CMC, where text-based communication is common, many context cues are eliminated entirely. It may be difficult to tell whether the person at the other computer is paying attention, understanding, agreeing, surprised, shocked, confused, or even in the room. Other context conditions, such as having to type a response, can frustrate participants and increase time pressure.

Several reviews and meta-analyses summarize the numerous CMC studies.[35] These do not always include the same studies, due to differences in the focus of the analyses and the criteria the authors use to select studies for inclusion. It is difficult to reach general conclusions, since many of the studies differ in factors shown to influence results: type of CMC (more often asynchronous e-mail than synchronous chat), task type (generate, choose, negotiate, execute), individual and group characteristics (often ad hoc groups with no established hierarchy), and research conditions (field or case study vs. laboratory experiment). Although some studies show contradictory findings in a number of areas, they are generally consistent in concluding that, relative to FTF groups, computer-mediated groups

- Have difficulty reaching consensus
- Have greater equality of participation
- Take longer to reach a decision
- Show differences in influence, particularly relating to status
- Exhibit lower inhibition
- Are more likely to be polarized

Several theories have been offered up to explain CMC groups' difficulty in reaching consensus.[36] Boland and Tenkasi (2001) suggest that perspective-making and -taking can be harder with CMC. Researchers investigating the process for arriving at consensus find that

---

[35] See Pinsonneault and Kraemer, 1989; Baltes et al., 2002; Bordia, 1997; Hollingshead and McGrath, 1995; Fjermestad, 2004; and McLeod, 1992.

[36] Hiltz and Turoff, 1978; Hiltz, Johnson, and Turoff, 1986; Hiltz and Johnson, 1989; Olaniran, 1994; Siegel et al., 1986; Straus and McGrath, 1994; Rice, 1984.

FTF groups generally show a gradual convergence, with later speakers being more likely to agree than to disagree with prior proposals. In CMC, this convergence does not happen as clearly.[37] George et al. (1990) theorize that the difficulty in reaching consensus might be related to CMC's equalization of participation. Past research has shown that consensus frequently follows a leader's push for his or her preferred solution.[38] Unlike FTF and AC collaborations, where leaders talk significantly more than other participants, in CMC, there is consistently more evenly distributed participation. All six of the studies on participation that McLeod included in her 1992 meta-analysis showed that CMC increased equality of participation. In Hollingshead and McGrath's 1995 review, 10 of 14 studies found more equal participation with CMC, and four found no difference. One might think that CMC "democratizes" the group, but in virtually all cases, the more equal participation was the result of *less* participation. Some people disengage entirely and "lurk," as indicated by 10 of 12 studies showing less overall participation in CMC.

Research has also shown that for many tasks, CMC groups take longer to reach a conclusion than do their FTF counterparts. Baltes et al.'s 2002 meta-analysis showed that CMC groups can take 4 to 10 times longer than FTF groups to make a decision or reach consensus, with the greatest difference occurring when the group is large or anonymous. This is consistent with McLeod's earlier meta-analysis, which showed that CMC groups take longer to complete their task than do FTF groups, as well as Pinsonneault and Kraemer's 1989 review and Hollingshead and McGrath's review, which found that 12 of 13 studies measuring task completion time found it to be longer in CMC. Reasons may include the inefficiency of typing versus talking; the higher CMC workload, with few resources available for attending to other group members;[39] lack of experience with the technology; and the lower degree of consensus in CMC collaboration.

---

[37] Hiltz, Johnson, and Turoff, 1986.

[38] See Rawlins, 1989.

[39] See Straus, 1996.

Interestingly, much more positive results have emerged in some instances. Bikson (1996) found that special support software called *groupware*, combined with facilitators and software operators called *technographers*, can decrease the time groups take to make decisions, although she warns that quick decisions are not necessarily the ones with the highest commitment to implementation.

If CMC groups often take longer to reach a conclusion and are less likely to reach consensus, what are the differences in discussion content? It is useful to break down the process into the four phases typically used in group-decisionmaking research: (1) prediscussion individual preferences, (2) group discussion, (3) group choice, and (4) postdiscussion individual preferences.[40] CMC affects the group discussion phase most, by altering the methods that individuals use to influence each other. Finholt and Sproull (1990) state that all groups will at some point be involved in influence attempts, and we can anticipate that virtual collaborators, who have responsibility for the group's outcome, are no exception. In FTF groups, cues such as poise, posture, intonation, facial expression, rate of speech, and confidence are used to establish credibility and influence others. How can CMC, which filters these cues, affect influence? It turns out that CMC participants make more explicit proposals, defer less to high-status members, and are less inhibited than are FTF collaborators.[41]

To explore CMC participants' tendency to make more explicit proposals, we compared the FTF and CMC processes seen in Rice's 1984 study. FTF foursomes who were given a problem to solve started out by analyzing it, whereas groups communicating by e-mail frequently started the discussion by proposing a solution. In some cases, all four members of the group suggested a solution before hearing anyone else's thoughts! This event was consistent with Hollingshead and McGrath's review, which found that all three relevant studies showed that CMC groups made more explicit proposals

---

[40] See, for example, Rice, 1984.

[41] See Siegel et al., 1986.

than FTF groups did. Again, typing skills may have contributed,[42] but leadership diffusion may also have been a factor. Hollingshead and McGrath's review showed that leaders are less likely to emerge in CMC groups, and leadership is more likely to be decentralized and less stable. In Dubrovsky, Kiesler, and Sethna's 1991 study of four-member decisionmaking groups, FTF groups showed the typical status inequalities: The high-status member dominated discussions, was more often a first advocate, and was more influential than the low-status members. When *the exact same groups* made comparable decisions using e-mail, status inequalities were reduced. Other studies have replicated this finding for other groups using e-mail[43] and electronic chat.[44]

One explanation for leadership diffusion comes from the evaluation of signals that leaders send in FTF situations. Leaders can emerge by dominating talk time or controlling the flow of the conversation. This is easy to detect in FTF groups but more difficult in the CMC context (Burgoon, 1978; Burgoon and Hale, 1988).

Unlike AC, where discussion tends to be more formal, CMC is often associated with disinhibition, or "behaviour that is characterised by an apparent reduction in concerns for self-presentation and judgement of others" (Joinson and Harris, 1998). There appears to be a spectrum of disinhibited behavior, the mildest forms being something as simple as dashing off a quick e-mail without regard to how it might be interpreted or being extremely candid.[45] Anyone who has experienced "flaming" in a discussion board or e-mail has noticed an extreme form of disinhibition, where people exhibit rude, impulsive behavior and express extreme views more often than they do in other

---

[42] Ibid.

[43] See Huff and King, 1988.

[44] See Kiesler and Sproull, 1992.

[45] See Sproull and Kiesler, 1991; Joinson and Harris, 1998; Hiltz, Turoff, and Johnson, 1989; Walther, 1997; Witmer, 1998; Robinson and West, 1992.

forums.[46] Hollingshead and McGrath found seven studies showing more disinhibited communication in CMC and three showing no difference. In one study, they found 102 uninhibited remarks in e-mail before consensus was reached vs. 12 in FTF discussions.[47] For one group, the electronic discussion was so heated that the participants had to be escorted separately from the building. In our own experience, flaming is more common now on Internet discussion boards than in professional communications, but misunderstandings can escalate when minor miscommunications go unchecked. Friedman and Currall (2002) theorize that conflicts are more likely to escalate in e-mail than in FTF or AC contexts. They note that e-mailers tend to bundle multiple arguments in single messages and often send and receive messages in relative isolation, "devoid of awareness of human sensibilities."

Why would people be less inhibited in CMC communications? Barefoot and Strickland (1982) theorize that the psychological distance imposed by CMC can allow a greater expression of emotions, especially negative emotions. Put another way, in a situation where social-context cues are strong, such as a meeting with a high-level decisionmaker, people's behavior tends to be relatively controlled and responsive to the status hierarchy. At a computer, where social cues are absent, people may forget that they are talking to another person, not a computer screen. Although Siegel et al. (1986) did not find a link between anonymity and disinhibition in CMC, the medium's *relative* anonymity may produce comparatively self-centered and unregulated behavior.[48] Uninhibited behavior in CMC may also be an attempt to reduce the number of proposals in order to reach consensus. Content analysis by Dubrovsky et al. showed that CMC's disinhibited statements were overwhelmingly angry, sanctioning statements, usually made against a group member for not conforming to

---

[46] Although flaming is now much less common in professional e-mail exchanges, as of the mid-1980s, it was common enough to motivate a thoughtful paper on e-mail etiquette (Shapiro and Anderson, 1985).

[47] See Dubrovsky, Kiesler, and Sethna, 1991.

[48] See Siegel et al., 1986.

the majority. Support for this finding comes from Weisband's 1992 study, where uninhibited remarks were negatively correlated with explicit proposals.

Recall that Bos's study of trust development in mediated groups found that VC and AC groups eventually obtained the same level of trust as FTF groups, albeit a more fragile and delayed trust. Bos found that CMC groups failed to achieve the same level of trust as groups using other media, a finding supported by Burgoon et al. (2003). Burgoon et al. also found that confederates assigned to be deceptive were believed *more than truth-tellers* in CMC contexts. Despite the tendency toward reduced trust, CMC may also reduce participants' ability to detect deception.

Many researchers have noted that participants will divulge more information with CMC than they would in FTF communication. One category of extra divulgence comes from strategic use, such as documenting a paper trail for later reference, using the "cc" feature to include others in communications between two people, or bypassing organizational hierarchy by sending e-mail directly to higher-level managers (Rocheleau, 2001). Another category of extra divulgence is what Witmer (1998) calls *risky CMC*—inappropriate disclosure of sensitive or personal information under the impression that CMC is private. This appears to be an extension of what Walther (1997) calls *hyperpersonal communication*, the tendency for CMC groups to exchange more intimate information about themselves than FTF groups do, which itself may be an extension of the tendency toward disinhibition discussed earlier. Ironically, CMC, with its written record, is less private and more easily distributed and traceable than FTF communication.

How does CMC affect the likelihood and level of conflict within the group? Poole, Holmes, and DeSanctis (1991) found that conflict reached higher levels in CMC than in FTF communication. A few researchers have found that CMC groups present higher levels of negative conflict (prolonging and escalating conflict, inflexibility, hostility, etc.) and lower levels of positive conflict (releasing tension, clarifying and reevaluating goals, creating new ideas, and so forth) than their FTF counterparts do.[49]

Outcome Differences Between Face-to-Face and Virtual Collaborations

Videoconference

As our evaluation framework in Figure 3.1 showed, collaborative outcomes consist of member attitudes (about the process, each other, and their output) and output characteristics. Despite early enthusiasm, attitudes about the VC experience are surprisingly negative.[50] Reports of phenomena such as users' instant dislike of people they have never seen before, self-consciousness about "being on TV," and jerkiness associated with low bandwidth slowed the adoption of VC capabilities. We suspect that some of the dissatisfaction with VC in the earlier studies comes from the relatively poor video and audio quality of old technology. As mentioned above, desktop VC, where individuals participate from their own offices, appears to have wider acceptance than more-formal VC arrangements. Even though the presentation quality of desktop VC is not much better than that of early conference-room VC, it is often preferred to AC.[51] Attitudes about group decisions in VC are similar to those in FTF meetings, with somewhat lower confidence in the VC decisions.[52]

Attitudes of VC participants toward each other are generally worse than those of FTF partners (Storck and Sproull, 1995), although media and task type can affect partner preference. Drolet and Morris (1995) found that VC negotiators tended to develop less rap-

---

[49] See Chidambaram, Bostrom, and Wynne, 1990; Zornoza et al., 1993; Zornoza, Ripoll, and Peiro, 2002; Straus, 1997.

[50] See Martin, 1977.

[51] From a 1982 survey by the Institute for the Future (reported in Egido, 1990) that showed that 50 percent of respondents would include video in the optimal office information system design.

[52] See Young, 1974b.

port, trust, and cooperation than did FTF negotiators. Purdy and Nye (2000) found VC negotiators more likely to compete rather than collaborate, and their collaboration efforts were less likely to be perceived by the other party, compared with FTF negotiators. In Williams' (1975b) experiment, 144 civil servants were assigned partners for a conversation that was either FTF, AC, or VC. In free-discussion conversations, FTF partners were strongly preferred to the partners in the other two modes. In a mildly competitive "priorities" discussion, VC partners were strongly preferred. Williams suggests that the additional interpersonal tension inherent in the more competitive task may have led participants to prefer a less intimate communication medium. This finding may be related to Manning, Goetz, and Street's 2000 finding that women can develop higher rapport over VC than in FTF communication. He theorized that women may feel more comfortable in the relative isolation of VC.

An intriguing explanation for the generally worse perception VC participants have of each other may come from a cognitive bias called the *fundamental attribution error* (Ross, 1977). This common bias occurs when one assumes that a person's actions are based on his or her disposition, rather than on the environmental situation. Conversely, people are more likely to attribute their own actions to situational factors than to dispositional factors (e.g., "He's late for the meeting because he's rude, but I'm late because I had to wait for a train"). Several researchers have found that mediated communication tends to exacerbate the fundamental attribution error.[53] Cramton (2002) explains that non-collocated groups have less common information about each other, about their situations, and about what they have done. Kelley and Michaela (1980) suggest that there is a "gradient of dispositional attribution as an inverse function of the total amount of information known about the other persons" (p. 477). Olson and Olson (2000) describe a meeting at which one group did not realize that the professor at the other end of the

[53] See Cramton and Wilson's 2002 series of three experiments; Abel's 1990 account of dispositional attribution during videoconferencing; Olson and Olson, 2000; Armstrong and Cole, 2002; Herbsleb and Grinter, 1999.

videoconference was missing an event important to him, the 15th anniversary of the liberation of Holland after World War II. The professor thought the executives "slow-paced" and "irritating," while the professor was seen as "increasingly curt."

What happens when part of a group is meeting FTF and other members are participating through VC or AC? Studies have demonstrated the formation of local coalitions, where participants show a significant bias toward supporting those at the same end of the telecommunications link.[54] AC participants also showed a significant bias against (i.e., the dissenter was more likely to be) the person on the other end of the telephone line.

Our own experience with VC and AC supports the notion of local coalitions. In fact, the medium almost encourages such coalitions; using the "mute" button on the voice channel automatically sends the message, "We don't want you guys to hear our planning." In addition, we note a tendency among AC groups to use visual signals to communicate with others in the same room, thereby excluding remote participants. Whether it's eye-rolling, "hurry up" motions with their hands, or facial expressions commenting on what was just heard, there seems to be an increased tendency to send "us vs. them" messages in AC.

Local coalitions can form beyond those in the same room. Three studies have shown a tendency in VC to form within-group coalitions rather than between-group coalitions.[55] For instance, VC participants from two different universities tended to side with participants from their own university, even if all individuals were connected by VC. Another study of dynamics between groups found significantly more communication breakdowns in VC and AC than in FTF.[56]

How are outcomes reached through VC different from those reached in FTF meetings? There is evidence that VC and AC produce more opinion shift in participants—that is, a greater difference be-

---

[54] See Williams, 1975a.

[55] See Strickland et al., 1978; Barefoot and Strickland, 1982.

[56] See Doerry, 1996.

tween their individual prediscussion and postdiscussion opinions—than occurs with FTF communication.[57] In a series of five experiments, Short, Williams, and Christie (1976) asked participants to rank order the seriousness of eight problems in Britain. The participants then discussed the issues and agreed on a final ranking and were asked their private opinions. All five experiments showed that opinion shift was much greater in VC and AC collaboration than in FTF.

Audioconference

Perception of participants toward each other has been found to be least positive in AC.[58] Recall that AC collaborators tend to form local coalitions with people in the same room, showing bias against the people on the other end of the line. Even when all participants are in different rooms, AC participants exchange less supportive communication (Stephenson, Ayling, and Rutter, 1976) and show a tendency for biased perceptions of each other.[59] In various experiments, members at the other end of an audio link were rated significantly lower on intelligence and sincerity than they were when they could see their partners.[60] In addition to perception biases, AC has been shown to cause participants to behave less generously and collaboratively (Morley and Stephenson, 1970; Purdy and Nye, 2000) and to be more aggressively competitive (Williams, 1977; Purdy and Nye, 2000).

Any collaboration has a socioemotional dimension and a task dimension.[61] Task-oriented activities include acquiring facts and in-

---

[57] See, for example, Meyers and Lamm, 1976; Reid, 1977.

[58] See Young, 1975; Williams, 1975b, 1977; Reid, 1977; Short, Williams, and Christie, 1976.

[59] Williams, 1975b, 1977; Reid, 1977; Short, Williams, and Christie, 1976.

[60] Williams, 1977; Reid, 1977; Short , Williams, and Christie, 1976.

[61] Various models have been proposed. Bales (1950) uses task vs. socioemotional; social identity theory uses interpersonal vs. normative; Deutsche and Gerard (1955) propose informational vs. normative. We believe Whitworth's (1998) cognitive three-process model (task, interpersonal, normative) has promise, although the older research does not map into this model directly. For our purposes, we have combined Whitworth's interpersonal and normative processes into Bales's socioemotional dimension.

formation, generating options, and objectively analyzing alternatives. Socioemotional activities include relating to others (forming and maintaining one-on-one relationships) and relating to the group (understanding norms, or expected behavior; forming a group identity; working toward consensus; and representing the group). Participants' perceptions of each other clearly affect the socioemotional side, but can they also affect the task at hand? Of McGrath's four task types (*generate, choose, negotiate,* and *execute*), we would expect *negotiate* and *choose* tasks to be most affected by participants' perceptions of each other.

Studies have shown that AC produces significantly less compromise and more total breakdowns in negotiation than does FTF communication.[62] In a series of studies of influence, Morley and Stephenson (1970) compared AC with FTF as participants negotiated an industrial dispute, one participant being assigned to the management side, the other to the union side. The participants were given information about the "facts" of the situation, which researchers had slanted so that one side or the other had a more objectively strong case. The studies found that the person with the stronger case was actually *more* successful with AC only than he was in FTF situations.[63] Evidently, the AC negotiators focused more on the facts than on the needs or styles of the negotiators. Morley and Stephenson (1970) hypothesized that

> the more formal the communications system the greater the emphasis will be placed on interparty rather than interpersonal aspects of the interaction. In the more formal audioconference, the more likely the settlement will be in accordance with the merits of the case.

They proposed that the communication medium could affect the balance between interparty and interpersonal concerns. This is supported by Short, Williams, and Christie's 1976 study, which

---

[62] See Short, 1972; Dorris and Kelley, 1972; Short, 1971.

[63] See Morley and Stephenson, 1969, 1970.

showed that when visual cues are removed, attention focuses instead on the verbal channel containing the interparty, task-oriented, cognitive material. Formality, defined in terms of the lack of social cues available,[64] is higher in AC and VC than in FTF communication (and indeed, experimenters have further manipulated formality by imposing rules prohibiting interruptions). Reid (1977) found that "the less formal FTF negotiation seems to place more emphasis on the human and reciprocal process of interpersonal communication, resulting in greater generosity and yielding by the side with the stronger case." This would imply that the effects of the communication medium are selective, affecting some types of negotiation that have interpersonal aspects, but not others.

Morley and Stephenson's seminal result, repeated by others, relies on the concept of "strength of case." To dissect this concept, Short, Williams, and Christie (1976) conducted an experiment in which one member of each pair negotiated a case of objective considerations, and the other negotiated a case based on opinion. The individual negotiating an opinion was allowed to choose his order of priorities among a list of nine potential areas of expenditure for a hypothetical company. The inverse order of priorities was then assigned to the other participant, who therefore constructed an objective argument. The results supported Morley and Stephenson's hypothesis: The person arguing the objective case was more successful via the more formal AC, but the person arguing his opinion was more successful with FTF. This finding, along with Morley and Stephenson's, is summarized in Figure 3.3.

These results can be interpreted as follows: When interpersonal cues are filtered via AC, participants focus on the impersonal aspects of the discussion. This was found to be true in Stephenson, Ayling, and Rutter's 1976 study. Participants with different views on management-labor relations were asked to discuss union-management negotiation problems either FTF or by AC. The researchers found

[64] See Morley and Stephenson, 1969.

Figure 3.3
Results of Morley and Stephenson and Short, Williams, and Christie

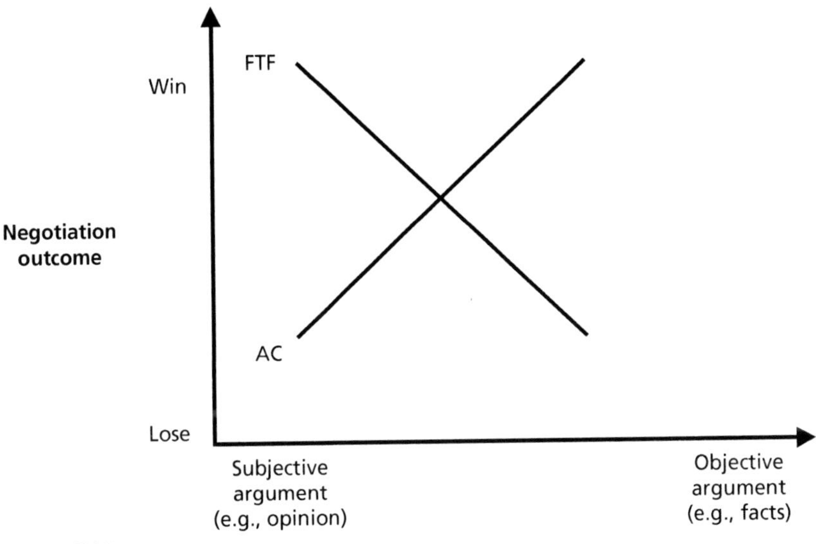

RAND *MG273-3.3*

that AC discussions were more task-oriented than FTF discussions and that participants spent less time maintaining relations. AC caused what the researchers called *depersonalization*, a feeling that other participants lacked personal qualities or individuality. Depersonalization was exhibited as reduced praise and more blame for the opponent (consistent with our later discussion on the fundamental attribution error being exacerbated), fewer self-references, and more overall disagreement. Content analysis showed purely informational interaction containing no reference to other conversational participants.

Another way AC's focus on impersonal aspects can affect outcomes is through the level of cooperation. Recall that social-dilemma experiments showed that VC and AC participants exhibit delayed and fragile trust. How does this affect cooperation? Researchers have run the prisoner's dilemma with pairs of subjects using different communication media.[65] Participants take on the role of prisoners, who can

[65] See LaPlante, 1971; Wichman, 1970.

either choose to cooperate (and get a lighter sentence if they confess) or not (and risk a heavier sentence if the other participant implicates them). Results showed significantly more cooperation in the VC mode than in AC. Again, the interpersonal aspect of cooperation has a greater effect in the less-formal VC.

Another outcome affected by AC is choice shift, or the difference between prediscussion individual choices and the group's choice. A series of 13 experiments using conflict tasks showed that AC discussion produces more choice shift than does FTF discussion.[66] This finding is somewhat counterintuitive, since FTF communication is richer, preferred, and more effective than AC, and one would predict greater choice shift in FTF collaboration. However, Morley and Stephenson's hypothesis argues that the FTF condition accentuates interpersonal and social aspects, thus distracting participants from exchanging and influencing opinions. Support for the distraction theory comes from a 2000 study in which participants using VC with degraded quality expressed more satisfaction with the process than those with higher-quality VC.[67] Another explanation for AC's higher choice shift comes from Salancik's 1977 study, which showed that people were less committed to their position when others could not see them. The results can be summarized as follows: It's harder to lose face when faces aren't visible.

In summary, whether the cause is more-negative participant impressions of each other, reduced occurrences of social interaction, or delayed and fragile trust, it is generally agreed that AC reduces interpersonal considerations. This does not seem to affect some tasks, e.g., information transmission or problem-solving, where AC and VC produce outcomes similar to those of FTF discussions.[68] However, AC can adversely affect the process and outcome for tasks involving person-perception, cooperation, trust, influence, and negotiation,

---

[66] See Reid, 1977; Short, Williams, and Christie, 1976.

[67] See Matarazzo and Sellen, 2000.

[68] For example, Reid, 1977; Harmon, 1995; Chapanis, 1975; Chapanis et al., 1972; Short, Williams, and Christie, 1976; Williams, 1977.

where the interpersonal aspects are more salient and more affected by the medium.

Computer-Mediated Communication

Satisfaction with the Group Process. Removing the voice and real-time video channel may significantly degrade participant satisfaction with CMC. Baltes et al.'s 2002 meta-analysis shows that collaborators are less satisfied with CMC when they are anonymous, have limited discussion time, or are in a large group. Other researchers have found equivocal results concerning member satisfaction with CMC,[69] possibly due to the task type, participants' status, or familiarity with computers.[70] Still other studies have found no relationship between media preference, prior computer experience, or gender (Ferris, 1995; Crowston and Kammerer, 1998). High-status participants are sometimes less satisfied with the CMC process,[71] where influence is more evenly distributed than it is in FTF. Wilson's 2003 study showed that FTF was perceived to be significantly more effective for persuasion tasks than CMC is and that participants use different strategies in different modes. Emotion was viewed as the most effective strategy for FTF and the least effective for CMC; logic was the second most effective strategy for FTF and the next-to-least effective for CMC; reward and punishment were viewed as the top persuasion strategies for CMC.

Attitude Toward Other Participants. As noted earlier, CMC participants may sometimes forget that they are talking to another person, not a computer screen. CMC reduces more social cues than even AC, and like AC, it has been shown to cause depersonalization, or the feeling that others lack personal qualities or individuality. From our earlier discussion of CMC's reduced trust and cooperation, it is not surprising that CMC groups have been shown to have reduced

---

[69] See Hollingshead and McGrath, 1995; Pinsonneault and Kraemer, 1989; Eveland and Bikson, 1992.

[70] See Fjermestad, 2004.

[71] See Brashers, Adkins, and Meyers, 1994.

peripheral awareness of their fellow team members (Bellotti and Bly, 1996; Benford et al., 1996). This may explain the finding by many researchers that CMC discussion is more impersonal and task-oriented than FTF discussion is,[72] although researchers also note that socializing via CMC takes more time and effort (Walther and Burgoon, 1992).

Much of the empirical research on CMC has used ad hoc groups of participants—typically students—without an established hierarchy. Although findings are sometimes replicated using existing work groups, little has been done to compare "one-shot" vs. ongoing teams and teams that already know each other. Three studies by Walther et al. showed that short-term groups using CMC tend to work in ways they usually do in FTF contacts, not adjusting as well to the medium as long-term groups do.[73] Over time, groups that have more than one contact with collaborators are more likely to exchange personal information and learn about each other, adjust to the asynchronous nature of e-mail, and work to maintain shared context and work flow.[74] Two studies by Walther et al. (2001) showed that short-term groups were more likely than long-term groups to blame their remote partners for their failures to adapt to CMC. They theorized that anticipated future contact made participants present themselves in ways that encouraged positive attribution to encourage others to work with them in the future. Harrison et al. (2003) did not find that anticipation of future interaction affected team performance, but they did find differences between teams that previously knew each other and those that did not. They found that teams connected with existing strong ties worked faster and created higher-quality products than those with no prior connections did. However, as time allowed familiarity to develop, the performance of the two types of teams

---

[72] Culnan and Markus, 1987; Hiltz, Johnson, and Turoff, 1986; Hiltz and Turoff, 1978; Hiltz, 1975; Hiltz, Turoff, and Johnson, 1989; Rice, 1984.

[73] See Walther, 1996; Walther, Anderson, and Park, 1994; Walther, Slovacek, and Tidwell, 2001.

[74] McGrath et al. (1993) found that CMC collaborators include more relational discussion over time.

converged. Harrison et al. proposed that "the time spent together may have provided a basis for coordination and division of labor and removed some of the process losses inherent in acquiring information about other team members" (2003, p. 660).

We speculate that the tendency to ignore interpersonal aspects when using CMC is typically stronger in teams that do not know each other or in one-shot teams than it is in groups with a preexisting close relationship. Nonetheless, in our own experience, we still find some evidence of depersonalization even among close colleagues, in that it is easy to dash off an e-mail without considering the context in which it will be received or to participate only minimally in CMC (or AC) while doing other work in parallel.

In contrast to FTF discussion, in CMC it is difficult to assess the recipient's mood, location, workload, or other contextual factors. Similarly, our colleagues read messages from us without considering the context in which we write them, which sometimes leads to miscommunications. In an early RAND report, Shapiro and Anderson (1985) proposed e-mail etiquette to reduce misunderstandings from media that "allow casual and formal messages to look superficially the same; that allow near-instantaneous, rather than reasoned, response; that don't permit feedback during the delivery of a message (as occurs in personal conversation); and that require modification to many old traditions of communication" (p. 2).

For some groups, CMC is generally believed to cause a state related to depersonalization called *deindividuation*, a broad category defined by Festinger, Pepitone, and Newcomb (1952) as the feeling of not being scrutinized or accountable when submerged into a group.[75] In fact, CMC's effect on deindividuation is so accepted that it is now used to manipulate deindividuation in laboratory experiments.[76] Deindividuation effects appear to be stronger in groups that are not in the same organization and are not linked professionally

---

[75] See Coleman, Paternite, and Sherman, 1999; Lea and Spears, 1991; Postmes, Spears, and Lea, 2002.

[76] See Spears, Lea, and Lee, 1990.

(Rocheleau, 2001). Postmes, Spears, and Lea (2002) summarized several researchers' findings as follows: "Electronic communication is sometimes seen as depersonalized or less individuated in the sense that the presence of individuals with whom one may interact is less visible or visible in a different way than in face-to-face interaction" (p. 3).[77] If CMC participants treat others more like objects than FTF participants do and consider themselves less scrutinized or accountable, how does that change the group's outcome?

Convergence to the Virtual Group's Norm. CMC may cause participants to behave more differently from FTF collaborators than VC or AC does. Several models have been proposed to explain the behavior of CMC groups, but the one that best fits the empirical results is Reicher, Spears, and Postmes's 1995 social-identity model of deindividuation effects (SIDE). The social identity (SI) portion of the SIDE model predicts that people's behavior is consistent with either their self-identity or the social identity of the group, depending on which is more salient. Various experiments have shown that manipulating the prominence of self-identity vs. social group identity (using mirrors, team names and symbols, etc.) causes people to behave according to the salient identity.[78] Which identity is more prominent in CMC collaborations? In anonymous groups, or even CMC where members are not absolutely anonymous, self-identity is less salient.[79] Even when CMC participants are identified to others, the medium offers what Postmes et al. call *relative anonymity*. Thus, the SIDE model predicts that in CMC, group identity will influence participants' behavior more than their individual identity will. Postmes, Spears, and Lea (1998) provide a useful review of the empirical support for the SIDE model.

The existing research does not show whether CMC increases the salience of group identity for experienced professionals who know

---

[77] Hiltz, Turoff, and Johnson, 1989; Jessup, Connolly, and Tansik, 1990; Kiesler, Siegel, and McGuire, 1984.

[78] See Postmes and Spears's 1998 meta-analysis and Ellemers, Spears, and Doosje's 1997 meta-analysis.

[79] Postmes and Spears, 1998.

each other well. We suspect that people might initially use their existing organizational identity—company, university, division—in CMC as a way of establishing credibility or a common bond, but do they act in accordance with group norms rather than individual norms more often in CMC? As we will see later, the in-group/out-group distinctions seem to be more common in mediated communications, but we must be cautious when applying the SIDE model to professionals in a tightly knit community.

The deindividuation effects (DE) portion of the SIDE model predicts two different impacts of CMC on social behavior. On the one hand, participants' view that they themselves are submerged into the group "offers the strategic liberty to ignore social pressures and unwanted influences" (Postmes, Spears, and Lea, 2002). On the other hand, participants have less information about others as individuals and are therefore more sensitive to information about the group. As Postmes et al. explain, "the relative anonymity associated with this medium provides a context in which individual differences between group members are sometimes less visible. As a result, the salience of group membership is likely to be accentuated"(2002, p. 4).

We have already seen evidence of this first impact of CMC—participants' ability to ignore social pressures and unwanted influence—in the findings of disinhibition; reduced participation; angry, sanctioning statements to get members to conform; and leadership diffusion (discussed in Chapter Two). The second impact—the accentuation of group membership—manifests itself in many ways. In a 1998 meta-analysis of 60 deindividuation studies, Postmes and Spears found that deindividuation increases responsiveness to specific in-group norms (expected behavior) and decreases the focus on personal identity, just as the SIDE model would predict. Reicher, Spears, and Postmes (1995) found higher normative influence in deindividuated groups. Studies by Hiltz, Johnson, and Turoff (1986) and Reid, Malinek, and Stott (1996) found that CMC groups produce up to four times more in-group–oriented messages (symbolism, solidarity, intergroup competition, etc.) than do FTF groups. Other studies support the finding that visual anonymity produces more conformance to the group's norm or more focus on in-group/out-group differentiation.[80] Spears et al. (2002) found that CMC groups were more likely to express opinions against an out-group than were FTF groups. Stereotypes against other groups may also be accentuated in CMC (Harmon, 1998; Thomson and Murachver, 2001); Douglas and McGarty (2002) found that when CMC participants were not anonymous, they used more stereotypical language to describe anonymous out-group targets.

CMC's focus on in-group/out-group differentiation seems related to the local coalitions formed around same-room considerations in AC and VC. Perhaps because CMC participants tend to be in different rooms, the "us-vs.-them" boundaries in CMC tend to be more "virtual." CMC's depersonalized context increases people's tendency to differentiate "us" from "them" on a variety of dimensions, from stereotypical bias such as gender to nationality and other social categories to attitudes to location.[81]

Location appears to be a common dividing line between "us" and "them." Researchers have found that when some of the partners are more geographically distributed than others, frustration is more often directed at the remote collaborators.[82] Walther et al. (2001) found instances in which collocated partners cast aspersions on their remote partners collectively ("What's wrong with those people?") and individually (describing them as lazy, irresponsible, or lacking commitment). Walther et al. attribute CMC groups' misattribution to their tendency to form especially potent in-group/out-group distinctions, combined with the frustration associated with adapting to CMC. Armstrong and Cole (2002) describe a "self-perpetuating feud" between subgroups that formed by location within a virtual software development team "with colleagues at one office site de-

---

[80] Postmes, Spears, and Lea, 2000 and in press; Spears et al., 2002; Lea, Spears, and DeGroot, 2001; Watt, Lea, and Spears, 2002; Postmes, Spears, and Cihangir, 2001; Wetherell, 1987.

[81] Postmes and Spears, 1998; Spears and Lea, 1994; Spears et al., 2000; Armstrong and Cole, 2002; Herbsleb and Grinter, 1999; Cramton, 2001.

[82] Burke et al., 1999; Cramton, 2001; Walther et al., 2001.

scribed as *us* and group members at distant sites labeled *them*." Similarly, Herbsleb and Grinter (1999) observed that subgroups formed by location in a virtual software development team in Germany and Britain. Finally, Cramton's (2001) study of dispersed teams under stress showed that "once in-group/out-group dynamics had arisen, there was a tendency for members of the subgroups to withhold information from each other. . . . Remote subgroups were described as 'lackadaisical,' 'aggressive,' and having an 'inferiority complex.'"

One recurring theme across all modes of virtual collaboration is that local coalitions are more likely to form in CMC than in FTF collaborations. In VC and AC, participants tend to show bias favoring the people in the same room. AC also biases participants away from remote participants. Location is a common theme in VC coalitions (e.g., universities) and CMC, with other dividing lines (social groups, gender, nationalities, attitudes) appearing in CMC. Figure 3.4 illustrates local coalition effects seen in VC, AC, and CMC.

The tendency of groups to differentiate "us" vs. "them" on attitudes has been noted by many researchers who have found that CMC reduces consensus and cohesiveness[83] and increases polarization, or the tendency for groups to become more extreme in their thinking following discussion.[84] Turner et al. (1987) explain CMC's increased polarization as follows: The CMC collaborators converge to the expected behavior (or norm) of their group—the in-group. The group differentiates itself from a hypothetical out-group with positions contrary to its own. This differentiation causes members to shift their opinions even further away from the out-group position, resulting in group decisions that are more extreme than the aggregate prediscussion choice. This choice-shift phenomenon occurs in FTF discussions but is more extreme in VC, AC, and CMC.[85] Although many

[83] See Straus, 1997; Hiltz, Johnson, and Turoff, 1986.

[84] See Lea and Spears, 1991; Spears, Lea, and Lee, 1990; Kiesler, Siegel, and McGuire, 1984; Siegel et al., 1986; Sia, Tan, and Wei, 2002.

[85] See McGuire, Kiesler, and Siegel, 1987; Kimura and Tsuzuki, 1998; Sproull, 1986; Kiesler and Sproull, 1992; Siegel et al., 1986; Weisband, 1992; Kiesler and Sproull, 1986.

Figure 3.4
Local Coalition Effects in Different Collaborative Modes

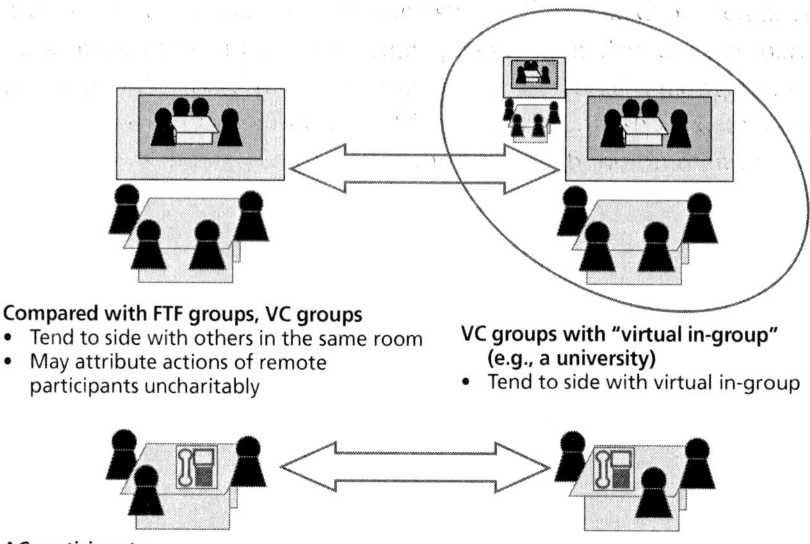

Compared with FTF groups, VC groups
- Tend to side with others in the same room
- May attribute actions of remote participants uncharitably

VC groups with "virtual in-group" (e.g., a university)
- Tend to side with virtual in-group

AC participants
- Tend to side with others in the same room and against those on the phone
- Judge remote participants as less intelligent and sincere than in FTF

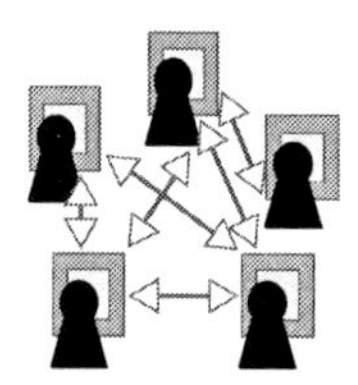

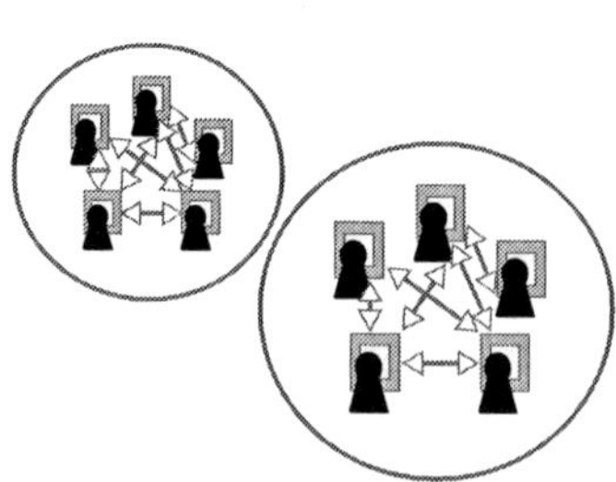

Compared with FTF groups, CMC groups more often have
- Increased polarization
- Tendencies to conform to group norm (expected behavior)
- Coherence bias
- Higher disinhibition
- Reduced group cohesiveness
- Tendency to make fundamental attribution error

CMC participants with a real or hypothetical outgroup
- Tend to side with others in the virtual in-group (e.g., a city)
- May attribute out-group uncharitably
- Tend to shift away from attitudes of the perceived out-group toward risky/extreme opinions

RAND *MG273-3.4*

researchers have noted that CMC increases group polarization, there is less literature showing that CMC increases choice shift. Still, the link between group polarization and choice shift is so strong that polarization is commonly evaluated using choice-shift measurements (Zuber, Crott, and Werner, 1992).

Effects of CMC on Outcomes. If CMC causes groups to interact differently, does it change their outcomes? There is evidence that in some cases, CMC improves group effectiveness. It generally increases task focus, which can improve outcomes for many types of collaborations. Group support software (GSS) can further focus team members on the task at hand by offering suggestions for steps in the process, organizing discussion into accessible threads, and providing tools such as anonymous voting, shared images, and models to help participants develop a common understanding of complex subjects. A number of studies have shown that for the *generate* task, CMC groups engaged in brainstorming activities produce a higher number of options with more novelty.[86] This effect, which is more pronounced in large groups, may be due to members' ability to type simultaneously without waiting for the "floor." Also, CMC participants—especially anonymous ones—are less apprehensive about how they will be evaluated than are FTF participants.[87]

In some cases, the wider variety of options generated in CMC can improve group judgment. Hiltz, Johnson, and Turoff (1986) found that CMC participants were more likely to "hold out" for their decisions, because there were fewer sanctions for people who deviated from the majority. And Hiltz, Johnson, and Turoff may be right; the intransigents contributed the highest-ranked solutions.

However, not all studies show that CMC improves brainstorming. Other factors, such as information overload, group size, and group type, may have a greater effect than communication mode. CMC's reduced cohesiveness has been shown to sometimes derail brainstorming, a task not typically sensitive to group cohesiveness. Stenmark (2002) found that in CMC brainstorming, participants

---

[86] See Fjermestad, 2004; Siau, 1995; Valacich, Dennis, and Connolly, 1994; Dennis and Valacich, 1993; and Gallupe, Bastianutti, and Cooper, 1991.

[87] See Silver, Cohen, and Crutchfield, 1994.

adopted an egocentric perspective, being more concerned with the reward system and getting credit for the good idea than their FTF counterparts were. Pinsonneault and Barki (1999) demonstrated that the best brainstorming comes not from CMC groups, but from nominal groups—those working alone and not interacting. Our experience is that in practice, it is difficult to prevent brainstorming groups from interacting, and CMC does tend to elicit novel suggestions.

CMC can also improve group effectiveness by mitigating certain cognitive biases. Social-group-membership bias occurs when people pay attention to, and are influenced more by, people like themselves. Bhappu, Griffith, and Northcraft's 1997 study showed that in CMC, even when the gender of participants was clearly identified, the gender-based social-group-membership bias disappeared. Similarly, McGuire, Kiesler, and Siegel (1987) found that when executives met FTF, men were five times more likely than women to make the first decision proposal. When the exact same groups met via computer, women made the first proposal as often as men. Another study showed that decisionmakers who rely on advice from others are distracted by nonvalid social cues, and this distraction is reduced in CMC.[88] Another bias that has been shown to be mitigated in CMC is availability bias, i.e., events that are more available to human memory are correspondingly judged as occurring more frequently or as being more important.[89] Finally, representativeness bias, the tendency to misestimate probabilities by not utilizing information sources such as base rate, was not mitigated by CMC alone but was reduced when a special problem-representation tool was used.[90]

Aside from these cases, however, CMC groups rarely, if ever, make better-quality decisions than FTF groups do (Baltes et al., 2002). CMC exacerbates some individual biases and also introduces new ones. Biases exacerbated by CMC include biased discussion, a

---

[88] See Hedlund, Ilgen, and Hollenbeck, 1998.

[89] See Benbasat and Lim, 2000.

[90] See Lim and Benbasat, 1997.

phenomenon where group discussion tends to focus on information that members already share before discussion and on information that supports, rather than opposes, the predominant sentiment within the group.[91] This may adversely affect decision quality; research has demonstrated that the degree to which unshared information is mentioned during a discussion is generally a good predictor of the ultimate decision quality.[92] Also, the sinister attribution bias, where people misattribute behavior of others to personal dispositions and overlook the influence of temporary, situational factors,[93] has been shown to be more likely to occur in electronic communication than in FTF.[94] This bias seems to be an extension of the fundamental attribution error discussed in Chapter Two, but here, attributions of the other person's behavior are assumed to be not only dispositional, but also diabolical.

Consistent with our earlier discussion about CMC's tendency to exacerbate out-group differentiation and stereotypes, studies have shown that CMC groups are more likely to fall victim to the fundamental attribution error than are FTF groups.[95] Researchers have found that the effort and slow pace of typing can cause participants to be more terse, leading them to be perceived as less polite (Daly, 1993; Walther, 1992a,b, 1994). Cramton and Wilson (2002) found that the distance between group members was statistically related to the way they attributed the behavior of others. "As the physical distance between team members increased, so did their tendency to make dispositional rather than situational attributions about each other." Cramton (2002) presents a case for CMC groups being subject to what Pettigrew (1979) calls the *ultimate attribution error*, where attributions are biased in favor of one's in-group and against a perceived

---

[91] See Hollingshead, 1998, 1996a,b; and Hightower and Sayeed, 1995, 1996.

[92] For example, Larson et al., 1998; Stasser and Stewart, 1992; Postmes, Spears, and Cihangir, 2001.

[93] See Ross, 1977.

[94] See Thompson and Nadler, 2002.

[95] See Cramton and Wilson, 2002, for three studies; Armstrong and Cole, 2002; Herbsleb and Grinter, 1999; Daly, 1993; Walther, 1992a,b, 1994; Walther et al., 2001; Abel, 1990.

out-group. Cramton cites factors that could contribute to the ultimate attribution error: "the development of strong subgroup identities by location, weak social integration of the dispersed team as a whole, a paucity of situational information concerning remote subgroups, information processing biased toward dispositional attribution concerning remote subgroups, and the challenging nature of collaboration under dispersed conditions, encouraging the creation of scapegoats." She cites evidence from her own studies, Armstrong and Cole (2002), and Herbsleb and Grinter (1999) to support her case.

A new bias that CMC introduces is coherence bias. Three studies have found that CMC participants detect interconnectedness and meaningful relationships even when no coherence is intended.[96] This may be a result of the higher levels of self-disclosure in CMC summarized by Joinson (2001) or Walther's hyperpersonal communications, mentioned earlier. Other biases introduced by CMC include temporal synchrony bias (the tendency for e-mailers to behave as if they are in a synchronous situation), the burned-bridge bias (where CMC collaborators engage in risky interpersonal behaviors in CMC that they would not engage in during FTF communication), and the squeaky-wheel bias (the tendency for CMC participants to adopt an aversive emotional style to achieve their goals).[97]

In negotiation tasks, CMC produces more errors than FTF discussion does. Arunachalam and Dilla's 1995 study of people negotiating the distribution of resources found that CMC negotiators showed lower judgment accuracy, had higher fixed-sum errors and incompatibility errors, obtained poorer outcomes (as measured by subsequent profits), and distributed resources less equitably than FTF groups did. This may be because CMC negotiators are not able to communicate their interests as clearly or easily as those negotiating FTF or because CMC negotiators' first offers tend to be more extreme than those of FTF negotiators.[98] For information-sharing tasks,

---

[96] See Cornelius and Boos, 2003.

[97] See Thompson and Nadler, 2002.

[98] See Shah, 1990.

CMC has been shown to produce fewer correct answers,[99] possibly because the electronic chat medium, with its simple text, limits members' ability to coordinate and verify information. For tasks with a time limit and those requiring a group consensus, CMC is less effective than FTF.[100]

Text-based communication such as CMC offers an opportunity to check and revise comments before they are sent. CMC also affords a shared and impartial transcript of the discussion that can be a useful reference, although one study showed that decisions made by CMC groups may be more often recorded incorrectly. Straus and McGrath (1994) found that CMC transcribers were more likely to willfully ignore other members' preferences.[101] Could this last finding explain the choice-shift phenomenon, implying that the group's decision was recorded incorrectly, away from the individual members' choice? It turns out that the correlation between the group's decision and the average of the individuals' postdiscussion opinions is high, indicating that on average, individual opinion shifts along with the group choice.[102]

Another difference in judgment outcomes between CMC and FTF groups may be an extension of an often-replicated phenomenon in FTF groups known as *risky shift*. Stoner discovered in 1961 that groups often make decisions that are more risky than any of the individuals' prediscussion judgments. As mentioned earlier, choice shift is higher for CMC, VC, and AC than it is for FTF. If choice shift is higher for CMC than for FTF, what about risky shift in CMC?

---

[99] See Graetz, Boyle, and Kimble, 1998.

[100] See Hollingshead and McGrath, 1995.

[101] Discrepancies between group decisions and recorded answers were found for 4 percent of FTF group responses and 24 percent of CMC group responses. FTF groups made one clerical error, and the remaining discrepancies were caused by recording the tentative agreement when time ran out. CMC groups made the same types of errors but also made new errors where the transcriber recorded a response with which one of the other members clearly disagreed (31 percent of discrepancies); recorded his or her own preference, disregarding the preference of other group members (6 percent); and recorded answers for issues the group had not addressed at all (13 percent).

[102] See Weisband, 1992.

Several experimental studies have shown that CMC has a tendency to make group decisions riskier, or more extreme, than FTF does.[103] In a typical assessment of risky shift, participants are given a written scenario with options that vary in risk and attractiveness (e.g., investment alternatives with high risk but potentially high payoffs, or an individual considering career choices with attractive but risky alternatives) and asked to indicate their prediscussion preference. The participants collaborate in small groups either by CMC or FTF and agree on an option. Frequently, a repeated-measures design is used, where participants repeat the exercise with a new scenario and a counterbalanced communication medium. Risky shift occurs when the group's choice is higher-risk than the individuals' average prediscussion choice. We use the term *riskier shift* to describe CMC's tendency to exacerbate FTF groups' risky-shift behavior.

It is debatable whether the choice shift seen in AC, VC, and CMC and the riskier shift seen in CMC should be considered judgment error. In some cases, riskier alternatives produce better outcomes. However, given that decisionmakers collaborate more often FTF and presumably have experienced the feedback and results of their FTF decisions more often, one could make the claim that they are more accustomed to the risk levels associated with FTF collaboration. Also, CMC participants most likely do not perceive their decisions to be riskier, making this effect invisible.

Several researchers have attempted to explain how CMC can shift the group's choice. We start with Stoner's 1961 discovery that FTF group discussion can cause people to advocate riskier courses of action than those advocated by people who did not participate in group discussion—the risky-shift effect. Later research showed that groups could shift their choices toward less-risky options, so the general term *choice shift* has been adopted. Deutsch and Gerard (1955) explained choice shift by defining two general processes of social influence in groups: informational and normative. The informational-influence explanation assumes that a variety of arguments are put

[103] See McGuire, Kiesler, and Siegel, 1987; Kimura and Tsuzuki, 1998; Sproull, 1986; Kiesler and Sproull, 1992; Siegel et al., 1986; Spears and Lea, 1994; and Weisband, 1992.

forth in group discussion and, to the extent that they are perceived as valid, will shift the choices of group members who had not previously considered these arguments.[104] Normative influence, in contrast, theorizes that participants will conform to the expectations of others (i.e., group norms). Choice shift is explained by members' motivation to equal or exceed the average group member on valued attributes.

Recently, researchers have used this distinction to explain CMC's effect on choice shift. As discussed earlier, the normative-influence (SIDE) model fits the empirical evidence better than the informational-influence explanation. So we start with CMC groups in a more deindividuated state than their FTF counterparts. How does this lead to the shift toward risky or more extreme choices? In the deindividuated state, individuals with visual or higher-level anonymity are less inhibited about suggesting extreme options and may be more likely to try to conform to the in-group norm. The group norm is not as obvious as it is in FTF communication, and consensus is more difficult, as we have shown earlier. Consequently, there is a tendency for participants to offer more explicit proposals and to use angry, sanctioning statements to encourage consensus. (Spears et al., 2002, found that disinhibited remarks are likely to be attempts to enforce the virtual group's norm.) Local coalitions are more likely to form in CMC than in FTF groups, perhaps to further define and enforce normative behavior. Out-groups are identified, either within the CMC group or outside, and the in-group attempts to differentiate itself by shifting its choice away from the out-group's position, resulting in a risky or extreme group choice.

Figure 3.5 synthesizes theories of Reicher, Cramton, Lea, Spears, Postmes, and Williams, among others, to describe how CMC can produce risky or extreme group choices.

CMC's deindividuation effects raise a concern about CMC increasing another phenomenon related to deindividuation—groupthink. Discovered by Janis in 1972, groupthink occurs when excessive concurrence-seeking overrides group members' motivation to realistically appraise alternative courses of action. Although the effects of communication media on groupthink have not been evaluated in the

---

[104] See McGuire, Kiesler, and Siegel, 1987.

Figure 3.5
Progression Toward Risky or Extreme Decisions in CMC

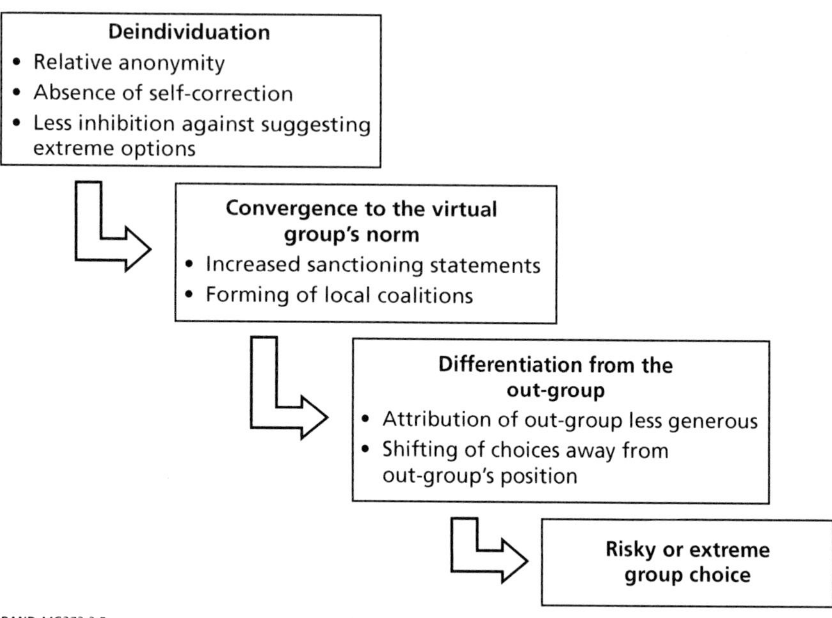

RAND *MG273-3.5*

literature, we can speculate on how groupthink might be affected by CMC.

Table 3.1 shows 't Hart's (1994) set of factors that contribute to groupthink. Some of these seem to be mitigated by CMC—tendencies toward anticipatory compliance with the leader's or high-status members' opinions/suggestions; cohesiveness; and concurrence-seeking—although the convergence to the virtual group's norm, described above, makes even this conclusion questionable.

How can groups be less cohesive, yet tend to converge to the virtual group's norm? The distinction between social and task cohesiveness, described by Carless and DePaola (2000), may provide a clue. Social cohesiveness pertains to the motivation of an individual to develop and nurture social relationships in a group. Task cohesiveness is the extent of motivation toward achieving the organization's goals and objectives. From the empirical findings of more task-oriented vs. social comments in CMC, we can infer that social cohesiveness is lower in CMC groups than in FTF groups. Straus (1997) found this to be the case and proposed that cohesiveness is lower in nonvisual communication because group members cannot identify similar

Table 3.1
Factors Contributing to Groupthink and Our Speculation on CMC Effects

Factor Contributing to Groupthink[a]	Potential CMC Effect
External pressures/threats to groups, notably conflict with other groups	Possibly exacerbated; biases against out-groups seen in CMC, along with fundamental attribution error and sinister-attribution bias
In-group's effectiveness in countering threats from out-groups	Possibly exacerbated; since CMC group cohesiveness is lower and out-groups are attributed less generously, this could be problematic
Increased group cohesiveness	Mitigated by CMC
Threats to self-esteem of decisionmakers; insecurity	Exacerbated by leader's relative discomfort with CMC and reduced influence
Perceived need for strong leadership	Possibly exacerbated by CMC's suppression of leadership emergence and less-stable leadership structure
Deindividuation tendencies	Exacerbated; CMC now used to manipulate deindividuation experimentally
Small group as decision unit	Not affected
Lack of established decisionmaking procedures	Exacerbated by CMC's more-frequent first-proposal offers, less-stable leadership structure, and intransigent, uninhibited behavior; group support software (GSS) may help.
Tendencies toward anticipatory compliance with leader's or high-status members' opinions/suggestions	Mitigated by CMC
Strong tendency for concurrence-seeking	Mitigated; concurrence-seeking not as high in CMC
Premature closure	Exacerbated; consensus takes longer; more angry, sanctioning statements and early proposals, and premature closure seen with group support software
Increased tendency to adopt high-risk alternatives	Exacerbated; choice shift toward extreme or risky options seen in CMC

NOTE: Shading of cells from light to dark indicates increasing likelihood of being exacerbated by CMC; unshaded cells indicate no effect.

[a]Identified in 't Hart, 1994.

demographic characteristics, a common feature of cohesive groups. If social cohesiveness is lower in CMC groups, we would speculate that the concurrence-seeking factor contributing to groupthink may be social rather than task-related and therefore mitigated by CMC.

Most of the other factors associated with groupthink seem to be exacerbated by CMC. Most notably, deindividuation and increased acceptance of high-risk alternatives seem to be more evident in CMC than in FTF collaborations. Other factors that may be intensified include conflict with other groups; insecurity of decisionmakers because of their reduced influence in CMC groups; lack of established decisionmaking procedures (except with decision support software); and premature closure. Miranda and Saunders (1995) found that teams using GSS identified more alternatives than FTF groups did but spent less time discussing those alternatives, indicating premature closure.

Other groupthink factors may be exacerbated by CMC; perceived need for strong leadership could be exacerbated by CMC's less central and stable leadership structure, and the in-group's effectiveness in countering threats from out-groups may be affected by CMC groups' tendency to draw in-group/out-group distinctions.

So what is our conclusion regarding the quality of group output? Empirical findings are mixed. Baltes' 2002 meta-analysis found that CMC groups rarely, if ever, made better decisions than their FTF counterparts. This may be because decisionmaking is a "convergent" task, requiring teams to understand the options put forth by others, to discuss the relative merits of the options, and to form consensus on the decision. Bikson (1996) notes that CMC is generally recognized as better-suited to "divergent" tasks such as idea generation, because of concurrency, editability, and anonymity—individuals can use CMC in parallel without waiting for the "floor"; participants can review and edit their comments before they hit "send"; the relative, and sometimes absolute, anonymity of CMC broadens participation because of reduced evaluation apprehension; and this anonymity allows more-objective review without being associated with an organization or hierarchy than is possible in a traditional meeting.

Under what circumstances does the higher-level structure offered by GSS improve team performance?[105] GSS is typically used by same-room/same-place groups[106] and thus is only tangentially related to virtual collaboration, where at least some of the participants are geographically dispersed. Nevertheless, GSS may have some applicability to our discussion, since teams could elect to get together FTF and use GSS for certain phases of a project. Dennis and Wixom (2002) provide a useful summary of the effect of five moderators on performance of virtual teams using GSS: (1) task, (2) tool, (3) group composition, (4) group size, and (5) facilitation. Dennis and Wixom define performance in terms of three factors: decision quality or number of ideas generated, time to complete the task, and participants' satisfaction with the process or outcomes. The results of their meta-analysis of 61 studies showed the following:

- Decision quality was lower for distributed GSS teams than for FTF teams.
- When groups used same-place/same-time GSS, they made better decisions and generated more ideas than groups meeting FTF did, but they took longer and had lower process satisfaction.
- Satisfaction was higher for GSS idea generation than for GSS decisionmaking.
- Teams using level 1 GSS (which includes provisions for simultaneous information input and anonymity) generated significantly more ideas than FTF groups but required more time.

---

[105] We note the mixed findings of various reviews and meta-analyses of GSS by McLeod, 1992; Benbasat and Lim, 1993; Benbasat, DeSanctis, and Nault, 1992; Fjermestad, 2004; Jarvenpaa and Shaw, 1998. We believe Dennis and Wixom (2002) have a useful organizing structure, as they evaluate the moderators' influence on processes and outcomes rather than simply the outcomes of teams using GSS.

[106] Valacich, Dennis, and Connolly (1994) define GSS as "an environment that contains a series of networked computer workstations that enable groups to meet face-to-face, with a computer-supported electronic communication channel used to support or replace verbal communication." Dennis and Wixom found that same-place/same-time GSS was used in more than 80 percent of the studies in their meta-analysis. In our definitional structure, same-place/same-time GSS would represent a hybrid between CMC and FTF communications.

- Teams using level 2 GSS (which includes information-analysis tools designed to organize, model, change, and rank information) made significantly better decisions and generated more ideas than FTF groups did, but they took more time and had lower process satisfaction.
- As group size increases, the extra time required for GSS idea generation diminishes.
- The presence of a GSS facilitator significantly improved decision quality and satisfaction with the process.

Thus we see that GSS is a promising development for large CMC groups, especially when they can use the tool FTF with a facilitator in the idea-generation phase. Dennis and Wixom envision virtual reality rooms, where distributed team members can reap many of the benefits of GSS from different locations. They also conclude that using GSS without a facilitator offers few benefits (except for idea generation) for inexperienced teams, and it may be prohibitively expensive to provide a facilitator for every team using GSS. However, we believe that advances in the technology, along with group experience and comfort with the tool, will eventually eliminate the need for external facilitation, although group members will then have to become effective in the facilitation function.

Although Dennis and Wixom found GSS to take longer than FTF meetings for decisionmaking and idea generation, several field studies have shown that GSS can decrease the calendar time and manpower required for other types of projects.[107] In Bikson's 1996 study of the World Bank's use of GSS to produce loan evaluations, participants perceived that groupware-supported meetings enabled them to exchange knowledge better and to have more impact on group outcomes than traditional meetings did. In addition, having an immediate, complete, shared, and impartial record of meeting processes and outcomes—a document generated from the groupware's transcript—improved timeliness, comprehensiveness, accuracy, and

[107] See Nunamaker et al., 1989; Dennis et al., 1990; Grohowski et al., 1990; Martz, Vogel, and Nunamaker, 1992; Post, 1992; Dyson, 1993; Nunamaker et al., 1997.

quality of the group's output. Similarly, Adkins, Burgoon, and Nunamaker (2002) report that the Air Force's GSS-supported strategic planning took less than half the time the participants had expected for FTF sessions, with no significant difference in commitment to implement the group's decision. The Boeing Corporation used virtual-collaboration GSS to develop an electronic mockup of its new 7E7 airliner, requiring one-sixth the number of designers and taking one-seventh the time of the old method, which developed a physical mockup.[108]

We conclude that the quality of a group's output depends on many factors, but we find that CMC generally produces better idea generation and documentation of the discussion, and FTF produces better cooperation, negotiation, and consensus. Tasks that require access to interpersonal information, such as forming a cohesive team, ensuring understanding, influencing, judging, bargaining, negotiating, resolving conflicts, and forming consensus, are most affected by CMC, which tends to produce more task focus. Many interpersonal effects of CMC may be obvious (tendencies for reduced consensus and status effects, increased disinhibition, deindividuation, depersonalization, and polarization), but others may be more subtle (local coalitions, shifts toward risky or extreme options, tendency to attribute others less generously). Since CMC seems to mitigate some cognitive and group biases, exacerbate others, and introduce new ones, and since several factors moderate its effects, further generalization about CMC's effectiveness for other tasks is not reliable.

Summary

The large number of study conditions, technology states, task types, group and individual variables, and mediating factors makes general conclusions about mediated communication difficult. For instance, research in FTF communication as well as CMC shows that group size affects participation and outcome, yet many studies use small,

---

[108] "SGI at Boeing," http://www.sgi.com/industries/manufacturing/aerospace/boeing.html.

three- to five-person groups. Researchers tend to use many small groups to maximize statistical power for results where the group is the unit of analysis. Palme (1995) claims that FTF meetings are most efficient with three to five people, whereas CMC groups (synchronous or asynchronous) are most efficient with larger groups. Palme finds 20 to 100+ people to be the group size where CMC efficiency surpasses that of FTF, while Dennis and Valacich (1993) find nine as a cross-over point. This would imply that studies comparing CMC and FTF groups of three to five people will tend to overestimate the disadvantages of CMC collaboration as group size increases.

Another concern arises when we try to generalize laboratory findings to real-world conditions. For example, researchers generally agree that there is less participation in CMC groups than there is in FTF groups. This is certainly consistent with our professional experience, where we often note reduced participation in AC, VC, and CMC. In laboratory conditions, we would expect participation to be artificially high due to the participants' desire to assist the researchers. If our conjecture is true, then the experimental results showing that CMC results in reduced participation are understating the effect.

In addition, groups assembled for experimental research may not resemble real-world groups. Older studies used small ad hoc groups of college students in a laboratory setting.[109] Many of those groups had no established hierarchy and were in an early or initial phase of team development, where teams typically do not focus on task.[110] It would be difficult to simulate ongoing professional relationships in a laboratory setting. As discussed earlier, when teams anticipate an ongoing relationship, they may share more interpersonal information.

Based on the deindividuation research, we would also expect to see more differences between FTF and CMC collaborations for tasks involving interpersonal considerations. Similarly, the type of task chosen for research also might not reflect real-world challenges. Many of the tasks examined were relatively simple judgment activities, yet it

[109] See Baltes et al., 2002.

[110] See Pinsonneault and Kraemer, 1989.

has been shown that the degree of complexity of the task is a moderating factor for CMC.[111] Finally, ethics considerations forbid putting research groups through stress levels that may be experienced by many real-world decisionmakers. We are encouraged by the numerous field studies that replicate the laboratory findings, using existing hierarchical work groups in their typical work setting.

A final issue is sampling bias. Research that is published typically has a bias toward statistically significant, hypothesis-confirming results. Here, we have attempted to summarize findings that have been replicated by researchers, and we have noted when only one study found conclusions we thought were significant.

Notwithstanding these concerns, when several studies with various combinations of conditions come to similar conclusions and those conclusions match our own experience, we feel more confident about the findings.

Our broad review of different communication modes reveals that some effects on context, process, and outcome are common across VC, AC, and CMC. All media provide a broader reach for participants and subject-matter experts, enabling them to make contact more quickly. Discussion tends to be more task-oriented and less social. However, visual cues are reduced, even in VC, making it harder to regulate and understand the conversation, detect understanding, and share visual aids. Despite the media's broader reach, participation is reduced, along with interpersonal factors such as perceptions of other collaborators, cohesiveness, persuasiveness, cooperation, and leadership emergence. Virtual groups are more likely to attribute others' behavior less generously, to dispositional rather than situational factors. There is a tendency for coalitions to form around people in the same room (for VC and AC) and around other "us vs. them" divisions such as universities (in VC), companies, countries, stereotypical groups, or attitudes (in CMC). Compared with FTF groups, virtual groups are less likely to reach agreement, and their decisions are shifted further away from the individuals' prediscussion choice.

---

[111] See Gallupe, DeSanctis, and Dickson, 1988; Bui and Sivasankaran, 1987.

Having identified the broad themes across VC, AC, and CMC, we now discuss the effects of specific modes. VC appears to be closest to FTF in process and outcome, but it reduces many visual cues. It increases participants' workload and may result in more orderly and polite conversation. Participation, though reduced, is more equal to FTF participation, and VC may be preferred for mildly competitive discussions. In general, VC participants are less satisfied with the process than FTF groups are. AC eliminates visual cues about remote participants. Compared with FTF, AC may exaggerate status relations, increase dominance by a few members, and increase influence for an objectively stronger case. In experiments, AC seems to reduce influence for subjective cases and to lower cooperation. AC produces greater choice shift than does FTF or VC. AC has been shown to make people perceive each other more negatively than do either VC, CMC, or FTF. VC and AC conditions are associated with more communication breakdowns than FTF meetings are.

CMC eliminates most visual and verbal communication. Unlike AC, it reduces status effects and domination by a few members. Accordingly, high-status participants, whose influence is reduced, may not be as satisfied with the process. Like AC and VC, CMC suppresses leadership emergence, but CMC makes leadership more likely to be decentralized and less stable. CMC makes consensus harder to reach and may increase the time to reach a decision or consensus. It can increase disinhibition, especially in (but not limited to) the anonymous condition, with honest, frank, and extreme comments seen more often in CMC. Cohesiveness and participation are lower in CMC than in FTF conditions. CMC's equalization and disinhibition effects can improve brainstorming, causing CMC groups to generate more options and more novel options than FTF groups do. Influence is shifted away from leaders, and CMC influences strategies toward more explicit proposals and the use of reward and punishment rather than emotion. Intransigents are more likely to hold out for minority views in CMC than in FTF groups. CMC generally increases deindividuation, depersonalization, polarization, and choice shift and is associated with groups making riskier or more-extreme decisions than FTF groups do. Just as VC and AC groups can form local coalitions with those in the same room or with the same affiliation, CMC groups tend to form local coalitions against virtual out-groups—those further away, in different cities, companies, or countries. Even within a CMC group, coalitions are more likely to form around attitude and against hypothetical out-groups. CMC can be used to mitigate some cognitive biases (social group membership, availability, representative biases), but it exacerbates other biases (biased discussion, fundamental attribution bias, sinister attribution biases). CMC also introduces some new biases (coherence, temporal synchrony, burned-bridge, and squeaky-wheel biases).

Table 3.2 summarizes the differences reported in the literature between FTF collaboration and VC, AC, and CMC collaboration, mapped into the framework of contextual variables, group process, and outcomes described earlier.

Table 3.2
Summary of Research Findings

Context Effects	Process Effects	Outcome Effects
Common to All Media		
Broader reach for participants Quicker response time Reduced social and other nonverbal cues Power and status harder to detect Visual aids harder to share	Conversation is • harder to regulate and understand • more task oriented, less social Participation reduced Leadership emergence suppressed Cohesiveness reduced	More choice shift More failures to reach agreement Interpersonal considerations reduced Fundamental attribution error exacerbated Cooperation reduced More likely to form "us vs. them" coalitions
VC-Specific		
Workload increased • members shift to simpler problem-solving strategies • reduced ability to raise counter-arguments	Conversation more orderly and polite More equal participation and influence Lower persuasiveness	Local coalitions form, biased toward those in room or in virtual group Lower decision confidence Less satisfied with process More communication breakdowns May be preferred for mildly competitive discussion Lower rapport

Continued

Table 3.2 (continued)

Context Effects	Process Effects	Outcome Effects
	AC-Specific	
Visual cues removed • increases task complexity	Conversation more formal Status effects increased: • more conversational dominance • participation less equally distributed	More choice shift than in VC Lower decision confidence Least positive ratings of other participants (viewed as less intelligent and sincere) Persuasion stronger for objective argument, weaker for subjective argument Local coalitions form, biased toward those in room, away from those on phone
	CMC-Specific	
Typing required No visual or auditory cues Increased task complexity Anonymity possible Reduced: • conformance pressure • inhibition • individual responsibility • effort Workload increased in synchronous CMC	Subcommittee structures more agile, elaborate; subcommittees share members Longer time to decision Consensus less likely Status inequalities reduced: • participation more equally distributed • high-status members less influential Leadership decentralized and less stable Lower inhibition Influence shifted away from leaders More explicit proposals, sometimes in first communication More sanctioning statements	Brainstorming produces more results More choice shift than with other media Shift toward risky or extreme options Some cognitive biases mitigated: social group membership bias, availability bias, representative bias (with GDSS) Some biases exacerbated: biased discussion, fundamental attribution error, sinister attribution bias New biases introduced: coherence bias, temporal synchrony, burned-bridge, squeaky-wheel biases Deindividuation increased Polarization increased Local coalitions form around real or hypothetical out-groups Cohesiveness lower Intransigents more likely to hold out for a minority view

CHAPTER FOUR

Mitigating Problems and Exploiting the Benefits of Mediated Communication

Broad Observations

In this chapter, we discuss techniques to mitigate adverse media effects while leveraging benefits the media can offer. It is generally held that certain task types lend themselves better than others to CMC. Information exchange seems to be almost as effective in CMC as in FTF, and brainstorming is improved with CMC. Conversely, FTF seems best for tasks requiring interpersonal exchange, such as when the decision requires complex thinking or negotiations, or when problems are ill defined.[1] Since FTF meetings are in person and improve cohesiveness, they are probably best for generating and checking commitment to a course of action. Figure 4.1 attempts to summarize several researchers' observations about media effectiveness as a function of the type of exchange. It is notional and simplified.

Tactics for Mitigating Problems

Many authors have recommended initial FTF interaction to create a relationship among group members, followed by mediated communi-

[1] See Kiesler and Sproull, 1992.

Figure 4.1
Notional Relationships Among Types of Communication

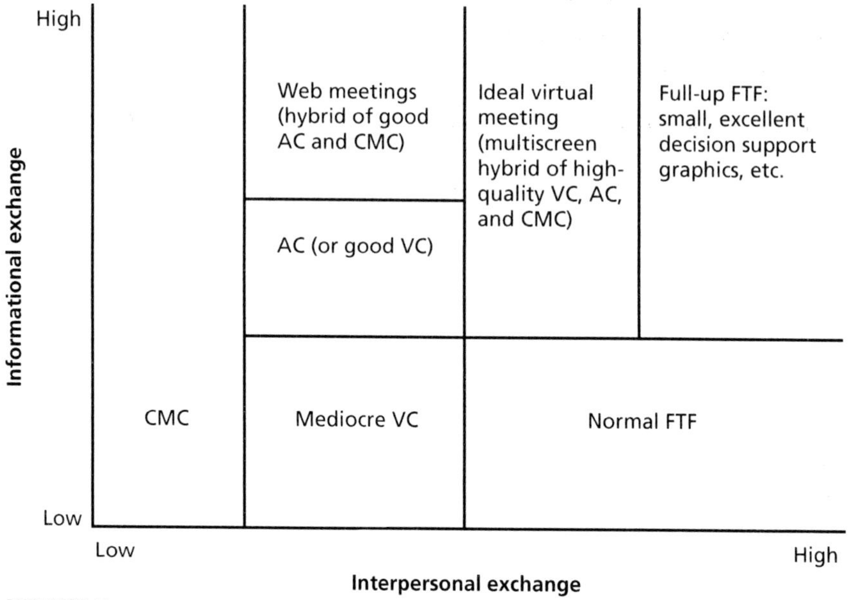

RAND *MG273-4.1*

cation to maintain the relationship.[2] Once the relationships, commitment, and cohesiveness are formed, groups could move toward mediated communication, with regular FTF collaboration to maintain working relationships.[3] 3-M recommends that its leaders "fly to build or repair trust." All of this is certainly confirmed by our personal experience. In the absence of this prior preparation, sharing biographical information, photos, and introductions can help virtual teams get to know each other. Moore et al.'s 1999 study of e-negotiators showed that brief personal disclosure over e-mail reduced the likelihood of impasse. Morris et al.'s 2002 study found that new groups collaborating by e-mail were significantly more likely to negotiate successfully if they had a brief telephone call before they

[2] See, for example, Nohria and Eccles, 1992; O'Hara-Devereaux and Johansen, 1994; Zielinski, 2000; Cramton, 2002; Zheng et al., 2002.

[3] See Brown, 1995; Slevin, et al., 1998; Kiesler and Sproull, 1992.

started e-mailing. The study showed that CMC participants who "schmoozed" once on the telephone developed more realistic goals, resulting in a broader range of possible outcomes, and developed better rapport with their electronic partners than those who communicated only electronically did. Another factor that increases participation in CMC was reported by Walther et al. (2001); in their study, groups were informed that their collaboration was either "one-shot" or an ongoing assignment. The findings, which replicate FTF group findings, indicated that groups anticipating a longer-term relationship had friendlier communication and exerted more effort.

The use of an AC facilitator has been shown to improve equality of participation and to help structure turn-taking and discussion flow.[4] Although choice shift and risky shift do not necessarily produce adverse consequences, being made aware of the shift through trained facilitation can bring the phenomena into group members' consciousness. In CMC, moderators have been used since the early days of the ARPAnet to keep order, hold votes, call FTF meetings for "theological conflicts," rein in uninhibited participants, and help keep discussion focused on the task.

At the anecdotal level, our personal experiences with facilitated discussion have been mixed to negative. Asking a "generic" facilitator to moderate a substantive discussion among researchers can be annoying and even insulting to the professionals involved. Better, in our view, is a trained facilitator with enough subject-matter knowledge to be credible, or a knowledgeable and respected individual acting as a reasonably neutral broker or moderator.

Several "rules for virtual groups" have been suggested by qualitative analyses of global and local virtual teams and have been shown to help group effectiveness. The suggestions include[5]

- More frequent communication than occurs in FTF groups

---

[4] See Rogelberg, O'Connor, and Sederburg, 2002.

[5] Walther et al. (2001) compiled this list, using research from Cramton, 2001; Flanagin, 1999; Jarvenpaa, Knoll, and Leidner, 1998; Jarvenpaa and Leidner, 1999; Mark, in press.

- Confirming receipt of messages
- Making sure that all members are included on all messages
- Early and continuous work on both organizing and making substantive contributions to the team's final project
- Substantially greater explicitness in questions, answers, agreements, and articulating expectations
- Earlier deadlines (more working ahead) and greater adherence to them

Other simple CMC work habits, such as quoting the person one is responding to in electronic chat, can greatly improve mutual understanding.[6] In addition, early agreement on ground rules (whether to vote, how long to spend on minority issues, etc.) may improve CMC effectiveness. Various software aids (e.g., the so-called "bozo filters") have come into use to filter out flamers, and other tools could be developed to give feedback to intransigents, informing them that their opinions are not being considered seriously.

Cramton (2002) found that a combination of techniques worked to mitigate the increased fundamental attribution error (the tendency to blame the person, not the situation) in CMC teams. When people violated their commitments, their remote partners were able to understand their situation if the violator (1) explained the situation that caused their failure and (2) expressed concern. Simply explaining the situation without expressing concern was not sufficient. Cramton also found that a rule of thumb predicted whether partners blamed violation of commitments on the violator or the situation: When one member of a team felt that he or she had done at least 50 percent more work than a partner had, that member tended to blame the partner for any violation of commitments. When the member perceived less than a 50 percent disparity, he or she acknowledged that the situation caused the problem.

---

[6] See Kreumpel, 2000; Vroman, and Kovacich, 2002.

Time can also improve the effectiveness of CMC collaboration in two ways: First, there is evidence that removing time constraints reduces outcome differences between CMC and FTF collaborations (Hiltz, Turoff, and Johnson, 1989; McLeod, 1992). Perhaps this is related to the finding that CMC groups that experience time pressure offer fewer affective statements ("I feel this way"), have harsher conflicts, and use poorer argumentation strategies (Reid et al., 1996, 1997). Also, some long-term studies have shown that experience with the medium improves CMC collaboration's effectiveness,[7] although Hiltz and Johnson (1989) found no significant change with time.

The benefits of CMC's anonymity—reduction of evaluation apprehension and pressure to conform—could be applied to other forms of communication, even FTF collaboration. Some arrangement for anonymous suggestions or anonymous voting might improve the breadth and quality of options discussed FTF or with other non-anonymous communications media.

Group decision support software (GDSS) adds another architectural layer to CMC by providing user interfaces and decision support technologies to improve problem representation, brainstorming, and solution. GDSS aims to "reduce the 'process loss' associated with disorganized activity, member dominance, social pressure, inhibition of expression, and other difficulties commonly encountered in groups and, at the same time, to increase efficiency and quality of resulting group decision" (Watson, DeSanctis, and Poole, 1988, p. 463). GDSS has been shown to improve task focus and decision quality over that attained by unstructured CMC.[8] Although GDSS is a relatively specialized technology, it has potential to mitigate adverse media effects in CMC and also to reduce more-common group effects. We envision GDSS providing various levels of service, depending on the amount of mitigation desired: (1) providing templates (e.g., for

[7] See Hollingshead, McGrath, and O'Connor, 1993; and McGrath and Hollingshead, 1994.

[8] See McLeod, 1992; Johnson, 1997.

announcements, agendas, options), (2) flagging potential issues for evaluation, (3) providing descriptive statistics, (4) suggesting actions, and (5) taking preprogrammed action to mitigate detected issues (e.g., enabling anonymous voting or taking breaks).

Whether implemented by GDSS or in the team's ground rules, there may be some advantage to structuring early CMC discussions to discourage early proposals. In one study of newly formed CMC groups, participants who listened to group discussion first and *then* advocated a position were more influential.[9] The data suggest that leaders emerge in groups because they are perceived as credible, and the way they gain that credibility is by listening to the group consensus.

Since CMC groups tend to focus more on the virtual group's norm than FTF groups do, leaders must establish clear expectations and goals for the group early, and they may be required to manage the group's behavior if problems arise. Postmes et al. (2001) have shown that some promising improvements in decision quality can be achieved by manipulating the CMC group's norm. When the group norm was primed to be one of independent and critical thought rather than one of consensus, groups made three times more correct decisions. We are encouraged by this result and suspect that many existing virtual groups are likely to have a culture of independent and critical thought, but it does raise questions regarding interactions between communication media and culture. Researchers have recently recognized the need to understand how individualistic vs. collectivist cultures collaborate via technology and to identify the media's effects on cultural misunderstandings.

Although we have developed a theoretical basis for groupthink being more likely in CMC, two factors contributing to groupthink could actually be mitigated by CMC. First, CMC decreases group cohesiveness and concurrence-seeking. And second, CMC reduces tendencies to comply with leaders' or high-status members' opinions or suggestions.

---

[9] See Weisband, 1992.

A Strategy for Choosing the Best Medium for Virtual Collaboration

If we overlay the informational vs. interpersonal model shown in Figure 2.1 with McGrath's group-task-type model shown in Figure 3.2, we can speculate on how communication media can affect typical tasks that a virtual team will perform. Figure 4.2 provides our summary considerations for choosing the communication medium to best fit the task at hand. The blocks on the left side of Figure 4.2 represent time-sequenced tasks performed by typical groups. The degree of shading represents the spectrum from tasks that may be adversely affected by virtual collaboration (darkest) to those that may benefit from virtual collaboration (lightest). For instance, forming a new team is best done FTF, where leaders can communicate goals and objectives and can get a sense of the group's understanding and commitment. In addition, early FTF communications help team members get to know their colleagues through informal meetings.

The right side of Figure 4.2 lists some challenges that participants may face as a function of the task and the communication mode chosen, along with some suggestions for overcoming the challenges. For instance, AC team members may form negative impressions of each other. Facilitators (or leaders performing the facilitation role) can partially mitigate this effect by expanding the introduction stage to include biographical information, hobbies, or other ice-breaking exercises. Similarly, facilitators should be aware of mediated groups' tendency to uncharitably attribute motivations of others, both in-group members and especially those in out-groups. It is tempting for leaders to introduce an "us vs. them" approach to overcome technology's damping effect on group cohesiveness, but facilitators may be surprised at the level of out-group bias that can develop naturally from CMC. Recall also that leaders emerge less clearly in CMC, so facilitators may have to explicitly establish subcommittee structures.

Figure 4.2
Strategy for Selecting the Best Medium for Virtual Collaboration

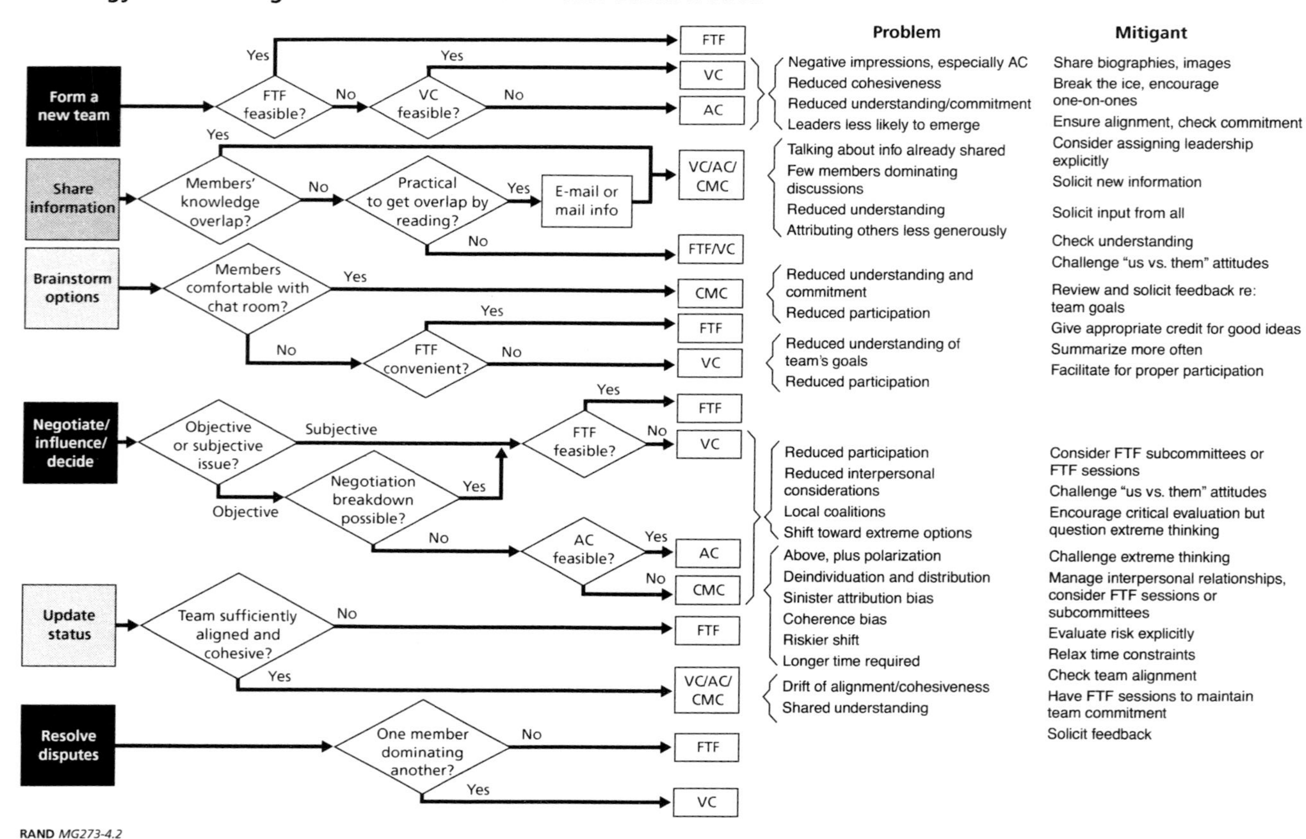

Information can be shared effectively via mediated communication if the members already have a common background or can obtain a common background by reading a document. Facilitators must be careful to avoid biased discussion, where participants focus on information that is already shared, and to ensure that all members are participating.

Brainstorming is a task that can be done more effectively via CMC than by other media, because participants can type while others are typing, and they tend to be less inhibited about offering unusual suggestions. Facilitators must be sure that CMC teams share an understanding and commitment to the project's goals before brainstorming options, and they must ensure that members' concerns about getting credit for good ideas are addressed.

Negotiating is difficult in mediated communication because of the media's effects on interpersonal considerations. Objective arguments are best done via AC, whereas subjective arguments are more effective in FTF presentations. In virtual teams—especially CMC—others' motives are attributed less charitably, polarization is more likely, shared understanding is less common, and participants may treat others less like people and more like objects. "Us vs. them" factions are more likely to appear in virtual teams, in the form of either local coalitions within the group or a tendency to attribute hypothetical out-groups less charitably. Facilitators must work hard to avoid negotiation breakdowns and should be aware that virtual teams are likely to choose more-risky or extreme options.

Status reporting can be done well via mediated communication, but leaders must work hard to ensure that the team is still aligned around a common objective. Facilitators should be aware that virtual team members may participate less than their FTF counterparts. Virtual teams may benefit from an occasional FTF session, where conversations can be more casual and cohesiveness can be boosted.

Dispute resolution is best done FTF, where interpersonal considerations are more salient. However, if one member is dominating another, VC may provide a less threatening environment for dealing with the dispute.

Having discussed both problems and possible ways to deal with them at this stage of technological development and understanding of the phenomena, we touch briefly in Chapter Five on where to go next.

CHAPTER FIVE

Where Next?

Keeping Up with Technology

As adoption of communications technology to improve productivity continues, our understanding of the technology's adverse effects must keep up. Some of the more obvious adverse effects, such as reduced participation and deindividuation, can be observed and mitigated. However, more-subtle effects may go unnoticed until the downstream consequences of group decisions appear. Problems may arise, for example, within the increasingly networked intelligence community that is being called for in light of the previously poor communications among agencies, and also in military operations, which already depend heavily on CMC, AC, and—at high levels—VC.

To illustrate how technology continues to introduce new phenomena, consider how new mobile-communication technology is changing group processes and outcomes in ways we can only begin to appreciate now. We all know some of this, as cell phones and other handheld devices can now access the Internet and can exchange e-mail, instant messages, and even pictures. But a wide range of users, from war-fighters to police to providers of tracking services to teenagers chatting with friends, are thirsty for mobile bandwidth. What will this increased bandwidth and use of wireless bring in? We are seeing one new decisionmaking phenomenon called a *smart mob*, in which people combine mobile short-message service (SMS) communication with computing capability to enhance real-time cooperation (Rheingold, 2002). A smart mob can have beneficial or destructive effects. In 2000, a million Filipino citizens toppled President Estrada, using text messages to organize a public demonstration. Also, the terrorists behind the Madrid bombings coordinated their efforts by cell phone and Internet.

Developing Better Conceptual Models

Most of the models used for mediated-communications research have their origins in FTF group communication. When the technology was first introduced, people used the new capabilities to mimic FTF communication. However, as technology such as GSS and mobile connectivity diverges further and further from traditional FTF communications, our existing conceptual models may get stretched beyond their design point. We have already seen evidence of this happening with cohesiveness research; the empirical findings routinely show that CMC is associated with lower cohesiveness but also with tendencies to conform to in-group norms and exchange more honest and intimate information. Although we have offered explanations for each of these observations, no holistic model exists to explain the seemingly contradictory observations.

Along those lines, we believe that new models of group phenomena might better predict and explain how they are affected by mediated communication. There is evidence that groupthink may be either exacerbated or mitigated by CMC, and the relationship between concurrence-seeking and mediated group behavior may give us clues that will help prevent groupthink.

Perhaps associated with groupthink, another group phenomenon, which we tentatively call *false sense of security*, may occur with CMC. Miranda and Saunder's 1995 finding of premature closure in CMC reminded us of other instances where things in virtual space happened too quickly—most notably, the faulty decision to deorbit the space shuttle Columbia was made using a combination of e-mail and AC, with engineers feeling that they were not fully understood (Kelly and Halvorson, 2003). Again, this example points to phenomena that may be unique to virtual collaborations and that therefore require new conceptual models.

Although it will take some time, we anticipate that new forms of models will be developed, perhaps drawing on concepts from other disciplines besides social psychology. We are starting to see evidence of this already, with contributions from evolutionary psychology (Kock, 2004), chaos theory (Maznevski and Chudoba, 2000), and software development (Ramash and Dennis, 2002). We suspect that in order to fully model CMC's complex interactions, a *systems-level* approach will be required, rather than the more traditional Newtonian approach of analyzing details of multiple interaction combinations. Connoisseurs of irony will note that researchers can use electronic databases, e-mail, and discussion boards to more effectively develop, critique, and validate new models of the CMC process.

Research Needed

Clearly, more research is needed to understand the causes and potential consequences of the media's effects on high-consequence decisions. Continued field research and case studies in specific environments will address some of our concerns regarding the generalizability of laboratory results and will surely reveal new, domain-specific effects. For example, networked intelligence analysts primed to "connect the dots" have, in addition to media-related influences, the well-known problem of motivated bias. And negotiators who already have reason to distrust each other by virtue of past experience or the complicated incentives affecting them may be even more subject to some of the adverse consequences of non-FTF discussion.

Also, we can speculate that substantial effects will occur in circumstances of clear importance that heretofore have been only weakly simulated in the laboratory. For example, we expect such effects for complex/ambiguous tasks where shared understanding is difficult, and for tasks with high consequences but also a high dependence on interpersonal considerations. Research on these matters should, in our view, include interviewing and on-site observation (even at top levels of government and business), because activities of this sort go on often, if not routinely, now but will continue to be resistant to laboratory study.

In addition, prospective, rigorous field-based research on virtual collaboration is necessary to overcome some of the limitations of laboratory studies mentioned in Chapter Three. We are encouraged by the increasing numbers of case studies appearing in the literature, but the complex interactions among factors affecting virtual collaboration require a more structured approach to make the findings more generalizable.

Much research is also needed on issues related to trust, which is essential to effective networked operations. Beyond the need to better understand purely human issues, such as how teenagers establish trust with strangers in electronic chats (and whether they learn how to maintain appropriate caution), there are new issues of a sort that make anyone with traditional command experience nervous, especially if that experience has included spot-checking to keep staff on their toes and to establish a basis for trust. In a networked world, leaders can no longer have confidence that their subordinates truly understand the "facts on the ground" about which they are reporting. Their information will itself have been obtained from disparate sources, many of which they will not be personally familiar with and over which they can enforce no quality control. Leaders will therefore be even less able than they were previously to have confidence in what they allegedly command and control.

Finally, we suggest further research on ways to teach people and organizations how best to facilitate virtual collaborations of different types. We suspect that any such teaching should be highly experiential and that pure textbook teaching will simply not suffice.

Training Needed

Most leaders acquire a sense of how group dynamics can help or hinder certain tasks when the group meets FTF. As we adopt new communications technology, management training must keep up.

Information-age managers should anticipate and deal with not only the observable effects of communications technology, such as polarization and reduced consensus, but also unseen effects such as choice shift, local coalitions, and disinhibition. Warkentin and Beranek (1999) point out that most CMC training deals with the technical side of the medium rather than the relational side. Their study showed that properly designed training can improve perceptions of cohesiveness and satisfaction with the CMC process, specifically with regard to trust, commitment, and frank expression between team members.

Opportunities

Although the focus of this report has largely been on identifying differences between mediated and FTF communications and on pointing out problems, we are in fact highly optimistic about the potential power of virtual collaboration. The capability of GSS is only starting to be exploited as we learn more about technology's impact on group behavior. The numerous GSS field studies show that when tailored to task and member characteristics, the technology can significantly improve the quality and timeliness of a variety of team outcomes. We anticipate the development and adoption of more generic GSS methodology to reduce the training and facilitation resources required.

In addition, there is promise that GSS may mitigate some of the challenges associated with group work. Moving problem-solving from the individual to the group level introduces process losses (such as socializing, domination, conformance pressure, etc.); the move to CMC may introduce new losses (riskier shift, polarization, in-group/out-group distinctions) but may recover some of the group process losses by equalizing participation and allowing anonymous input. Just as a smart mob might exchange anonymous text messages and discover that the emperor has no clothes, we may discover capabilities for major advances in teamwork beyond what can be accomplished today.

Bibliography

Abel, M. J., "Experiences in an exploratory distributed organization," in J. Galegher and R. Kraut (eds.), *Intellectual Teamwork: Social and Technological Foundations of Cooperative Work*, Hillsdale, NJ: Lawrence Erlbaum Associates, 1990, pp. 111–146.

Adkins, Mark, Michael Burgoon, and Jay F. Nunamaker, Jr., "Using group support systems for strategic planning with the United States Air Force," *Decision Support Systems*, Vol. 34, 2002, pp. 315–337.

Adrienssen, J.H.E., and J. M. van der Velden, "Teamwork supported by interaction technology: The beginning of an integrated theory," *European Work and Organizational Psychologist*, Vol. 3, Issue 2, 1993, pp. 129–144.

Anderson, A. H., A. Newlands, and J. Mullin, "Impact of video-mediated communication on simulated service encounters," *Interacting with Computers*, Vol. 8, No. 2, June 1996, pp. 193–206.

Armstrong, D. J., and P. Cole, "Managing distances and differences in geographically distributed work groups," in P. Hinds and S. Kiesler (eds.), *Distributed Work*, Cambridge: MIT Press, 2002, pp. 167–186.

Arunachalam, V., and W. N. Dilla, "Judgment accuracy and outcomes in negotiation: A causal modeling analysis of decision-aiding effects," *Organizational Behavior and Human Decision Processes*, Vol. 61, No. 3, March 1995, pp. 289–304.

Bales, R. F., *Interaction Process Analysis: A Method for the Study of Small Groups*, Cambridge, MA: Addison-Wesley, 1950, as cited in France, Anderson, and Gardner, 2001.

Bales, R. F., F. L. Strodtbeck, T. M. Mills, and M. E. Roseborough, "Chat communication in small groups," *American Sociological Review*, Vol. 16, 1951, pp. 461–468, as cited in France, Anderson, and Gardner, 2001.

Baltes, B. B., M. W. Dickson, M. P. Sherman, C. C. Bauer, and J. S. LaGanke, "Computer-mediated communication and group decision making: A meta-analysis," *Organizational Behavior and Human Decision Processes*, Vol. 87, No. 1, 2002, pp. 156–179.

Barefoot, J. C., and L. H. Strickland, "Conflict and dominance in television-mediated interactions," *Human Relations*, Vol. 35, No. 7, 1982, pp. 559–566.

Bellotti, V., and S. Bly, "Walking away from the desktop computer: Distributed collaboration and mobility in a product design team," in M. Ackerman (ed.), *Proceedings of the Conference on Computer Supported Cooperative Work*, Boston, MA: ACM Press, 1996, pp. 209–218, as cited in Cramton and Wilson, 2002.

Benbasat, I., G. DeSanctis, and B. Nault, "Empirical research in managerial support systems: A review and assessment," in C. W. Holsapple and A. Whinston (eds.), *Recent Developments in Decision Support Systems*, Springer-Verlag, 1992, pp. 383–437.

Benbasat, I., and L. Lim, "The effects of group support system on group meeting process and outcomes: A meta-analysis," *Small Group Research*, November, 1993, pp. 430–462.

Benbasat, I., and L. Lim, "Information technology support for debiasing group judgments: An empirical evaluation," *Organizational Behavior and Human Decision Processes*, Vol. 83, No. 1, September 2000, pp. 167–183.

Benford, S., C. Brown, G. Reynard, and C. Greenlaugh, "Shared spaces: transportation, artificiality, and spaciality," in M. Ackerman (ed.), *Proceedings of the Conference on Computer Supported Cooperative Work*, Boston, MA: ACM Press, 1996, pp. 77–85, as cited in Cramton and Wilson, 2002.

Berger, J. M., M. H. Fisek, R. Z. Norman, and M. Zelditch, Jr., *Status Characteristics and Social Interaction: An Expectation-States Approach*, New York: Elsevier, 1977, as cited in France, Anderson, and Gardner, 2001.

Berkowitz, B., "Failing to keep up with the information revolution," *Studies in Intelligence*, Vol. 47, No. 1, May 6, 2003.

Bhappu, A., T. Griffith, and G. Northcraft, "Media effects and communication bias in diverse groups," *Organizational Behavior and Human Decision Processes*, Vol. 70, No. 3, 1997, pp. 199–205.

Bikson, T., "Groupware at the World Bank," in C. Ciborra (ed.), *Groupware and Teamwork,* New York: Wiley, 1996, pp. 145–183.

Boland, R. J., Jr., and R. V. Tenkasi, "Communication and collaboration in distributed cognition," in G. M. Olson, T. W. Malone, and J. B. Smith (eds.), *Coordination Theory and Collaboration Technology*, Mahwah, NJ: Lawrence Erlbaum Associates, 2001, pp. 51–66.

Bordia, P., "Face-to-face versus computer-mediated communication: A synthesis of the experimental literature," *Journal of Business Communication*, Vol. 34, 1997, pp. 99–120.

Bos, N., J. S. Olson, D. Gergle, G. M. Olson, and Z. Wright, "Effects of four computer-mediated communications channels on trust development," *Proceedings of the CHI 2002 Conference on Human Factors in Computing Systems*, Minneapolis, MN, April 20–25, 2002, pp. 135–140.

Brashers, D. E., M. Adkins, and R. A. Meyers, "Argumentation in computer-mediated group decision making," in Lawrence R. Frey (ed.), *Group Communication in Context: Studies of Natural Groups*, Hillsdale, NJ: Lawrence Erlbaum Associates, 1994, pp. 263–282.

Brown, S. E., "The impact of electronic mail usage on the influence processes in geographically dispersed decision-making groups," *Dissertation Abstracts International, Section A: Humanities and Social Sciences*, Vol. 56, No. 6-A, December 1995, p. 2421.

Bui, T., and T. R. Sivasankaran, "GDSS use under conditions of group task complexity," Monterey, CA: U.S. Naval Postgraduate School, 1987, as cited in Pinsonneault and Kraemer, 1989.

Burgoon, J. K., "A communication model of personal space violations: Explication and an initial test," *Human Communication Research*, Vol. 4, 1978, pp. 129–142.

Burgoon, J. K., and J. Hale, "Nonverbal expectancy violations: Model elaboration and application to immediacy behaviors," *Communication Monographs,* Vol. 55, 1988, pp. 58–79.

Burgoon, J. K., G. M. Stoner, J. A. Bonito, and N. E. Dunbar, "Trust and deception in mediated communication," *Proceedings of the Hawaii International Conference in Systems Science,* Kona, HI, 2003.

Burke, K., K. Aytes, L. Chidambaram, and J. J. Johnson, "A study of partially distributed work groups: The impact of media, location, and time perceptions and performance," *Small Group Research,* Vol. 30, No. 4, 1999, pp. 453–490.

Carless, S. A., and C. DePaola, "The measurement of cohesion in work teams," *Small Group Research,* Vol. 31, 2000, pp. 71–88.

Carletta, J., S. Garrod, and H. Fraser-Krauss, "Placement of authority and communication patterns in workplace groups: The consequences for innovation," *Small Group Research*, Vol. 29, 1998, pp. 531–559.

Chapanis, A., "Interactive human communication," *Scientific American*, Vol. 232, 1975, pp. 34–42.

Chapanis, A., R. Ochsman, R. Parish, and G. Weeks, "Studies in interactive communication: The effects of four modes on the behaviour of teams during cooperative problem solving," *Human Factors*, Vol. 14, 1972, pp. 487–509.

Chapman, D. S., "Modeling job applicant decision processes: Integrating applicant reactions to selection procedures into the critical contact framework of recruiting," *Dissertation Abstracts International, Section B: The Sciences and Engineering*, Vol. 61, No. 10-B, May 2001, p. 5605.

Chidambaram, L., R. Bostrom, and B. Wynne, "A longitudinal study of the impact of group decision support systems on group development," *Journal of Management Information Systems*, Vol. 7, 1990, pp. 7–25.

Clark, H. H., and D. Wilkes-Gibb, "Referring as a collaborative process," *Cognition*, Vol. 22, 1986, pp. 10–39, as cited in France, Anderson, and Gardner, 2001.

Coleman, L. H., C. E. Paternite, and R. C. Sherman, "A reexamination of deindividuation in synchronous computer-mediated communication," *Computers in Human Behavior*, Vol. 15, No. 1, January 1999, pp. 51–65.

Cornelius, C., and M. Boos, "Enhancing mutual understanding in synchronous computer-mediated communication by training: Trade-offs in judgmental tasks," *Communication Research*, Vol. 30, No. 2, 2003, pp. 147–177.

Cramton, C. D., "The mutual knowledge problem and its consequences for dispersed collaboration," *Organization Science*, Vol. 12, No. 3, 2001, pp. 346–371.

Cramton, C. D., "Attribution in distributed work groups," in P. Hinds and S. Kiesler (eds.), *Distributed Work*, Cambridge, MA: MIT Press, 2002, pp. 191–212.

Cramton, C., and J. M. Wilson, "Explanation and judgment in distributed groups: An interactional justice perspective," paper presented at the Academy of Management annual meeting, Denver, CO, August 2002.

Crowston, K., and E. E. Kammerer, "Coordination and collective mind in software requirements," *IBM Systems Journal*, Vol. 37, Issue 2, 1998, pp. 227–247.

Culnan, M. J., and M. L. Markus, "Information technologies," in F. M. Jablin, L. L. Putman, K. H. Roberts, and L.W. Porter (eds.), *Handbook of Organizational Communication: An Interdisciplinary Perspective*, Thousand Oaks, CA: Sage Publications, 1987.

Daft, R. L., and R. H. Lengel, "Organizational information requirements, media richness, and structural design," *Management Science*, Vol. 32, 1986, pp. 554–571.

Daly, B., "The influence of face to face versus computer mediated communication channels on collective induction," *Accounting, Management, and Information Technologies,* Vol. 3, 1993, pp. 1–22, as cited in Whittaker, 2003.

Dennis, A. R., A. R. Heminger, J. F. Nunamaker, and D. R. Vogel, "Bringing automated support to large groups: the Burr-Brown experience," *Information and Management,* Vol. 18, 1990, pp. 111–121, as cited in Nunamaker et al., 1989.

Dennis, A. R., F. G. Joey, L. M. Jessup, J. F. Nunamaker, Jr., et al., "Information technology to support electronic meetings," *MIS Quarterly*, Vol. 12, No. 4, 1988, pp. 591–624.

Dennis, A. R., and J. S. Valacich, "Computer brainstorms: More heads are better than one," *Journal of Applied Psychology*, Vol. 78, No. 4, August 1993, pp. 531–537.

Dennis, A. R., and B. H. Wixom, "Investigating the moderators of group support system use," *Journal of Management Information Systems,* Vol. 18, No. 3, Winter 2002, pp. 235–258.

DeSanctis, G., and B. Gallupe, "A foundation for the study of group decision support systems," *Management Science,* Vol. 33, No. 5, 1987, pp. 589–609.

Deutsch, M., and H. B. Gerard, "A study of normative and informational social influences upon individual judgment," *Journal of Abnormal and Social Psychology,* Vol. 55, 1955, pp. 629–636, as cited in McGuire, Kiesler, and Siegel, 1987.

Doerry, E., "An empirical comparison of copresent and technology-mediated interaction based on communication breakdown," *Dissertation Abstracts International, Section B: The Sciences and Engineering,* Vol. 57, No. 1-B, July 1996, p. 453.

Dorris, J. W., and H. H. Kelley, *The Effects on Bargaining of Problem Difficulty, Mode of Interaction, and Initial Orientations,* Amherst, MA: University of Massachusetts, 1972.

Douglas, K. M., and C. McGarty, "Internet identifiability and beyond: A model of the effects of identifiability on communicative behavior," *Group Dynamics,* Vol. 6, No. 1, March 2002, pp. 17–26.

Drolet, A. L., and M. W. Morris, "Communication media and interpersonal trust in conflicts: The role of rapport and synchrony of nonverbal behavior," paper presented at the Academy of Management meeting, Vancouver, Canada, 1995, as cited in Purdy and Nye, 2000.

Dubrovsky, V. J., S. Kiesler, and B. Sethna, "The equalization phenomenon: Status effects in computer-mediated and face-to-face decision-making groups," *Human-Computer Interaction,* Vol. 6, 1991, pp. 119–146.

Duncan, S., "Some signals and rules for taking speaking turns in conversation," *Journal of Personality and Social Psychology,* Vol. 23, 1972, pp. 283–292, as cited in Williams, 1975a.

Dyson, E., "What IBM needs is a little team focus," *Computerworld,* Vol. 27, April 1993, as cited in Nunamaker, 1997.

Egido, C., "Teleconferencing as a technology to support cooperative work: Its possibilities and limitations," in J. Galegher and R. Kraut (eds.),

Intellectual Teamwork: Social and Technological Foundations of Co-operative Work, Hillsdale, NJ: Lawrence Erlbaum Associates, 1990, pp. 351–371.

Ellemers, N., R. Spears, and B. Doosje, "Sticking together or falling apart: Group identification as a psychological determinant of group commitment versus individual mobility," *Journal of Personality and Social Psychology*, Vol. 72, 1997, pp. 617–626.

Eveland, J. D., and T. K. Bikson, "Work group structures and computer support: A field experiment," in R. Baecher (ed.), *Readings in Groupware and Computer-Supported Cooperative Work: Assisting Human-Human Collaboration*, San Francisco, CA: Morgan Kauffman, 1992.

Ferran-Urdaneta, C., "The effects of videoconferencing on persuasion," *Dissertation Abstracts International, Section A: Humanities and Social Sciences*, Vol. 61, No. 8-A, March 2001, p. 3239.

Ferris, S. T., "An investigation of a role of computer-mediated communication as a media choice in the facilitation of task performance in small groups," *Dissertation Abstracts International, Section A: Humanities and Social Sciences,* Vol. 56, No. 5-A, November 1995, p. 1581.

Festinger, L., A. Pepitone, and T. Newcomb, "Some consequences of de-individuation in a group," *Journal of Abnormal and Social Psychology*, Vol. 47, 1952, pp. 382–389.

Finholt, T., and L. S. Sproull, "Electronic groups at work," *Organization Science*, Vol. 1, No. 1, 1990, pp. 41–64.

Finn, K. E., A. J. Sellen, and S. B. Wilbur (eds.), *Video-Mediated Communication*, Hillsdale, NJ: Lawrence Erlbaum Associates, 1997.

Fjermestad, J., "An analysis of communication mode in group support systems research," *Decision Support Systems*, Vol. 37, Issue 2, 2004, pp. 239–263.

Fjermestad, J., S. R. Hiltz, and M. Turoff, "An integrated framework for the study of group decision support systems," *Proceedings of the 26th Hawaii International Conference on System Sciences*, 1993, pp. 179–188, as cited in Whitworth, 1998.

Flanagin, A. J.. "Theoretical and pedagogical issues in computer mediated interaction and instruction: Lessons from the use of a collaborative instructional technology" *Electronic Journal of Communication*, Vol. 9, No. 1, 1999, as cited in Walther et al., 2001.

France, E. F., A. H. Anderson, and M. Gardner, "The impact of status and audio conferencing technology on business meetings," *International Journal of Human-Computer Studies*, Vol. 54, No. 6, 2001, pp. 857–876.

Friedman, R. A., and S. C. Currall, "E-mail escalation: Dispute exacerbating elements of electronic communication," *Proceedings of the International Association for Conflict Management's 15th Annual Conference*, Salt Lake City, UT, 2002.

Gale, S., "Human aspects of interactive multimedia communication," *Interacting with Computers*, Vol. 2, 1990, pp. 175–189, as cited in Isaacs and Tang, 1994.

Gallupe, R. B., L. M. Bastianutti, and W. H. Cooper, "Unlocking brainstorms," *Journal of Applied Psychology*, Vol. 76, 1991, pp. 137–142.

Gallupe, R. B., G. DeSanctis, and G. Dickson, "Computer-based support for group problem finding: An experimental investigation," *MIS Quarterly*, Vol. 12, No. 2, 1988, pp. 277–296, as cited in Pinsonneault and Kraemer, 1989.

George, J. F., G. K. Easton, J. F. Nunamaker, Jr., and G. B. Northcraft, "A study of collaborative group work with and without computer-based support," *Information Systems Research*, Vol. 1, 1990, pp. 440–457.

Gibson, C., and S. Cohen (eds.), *Virtual Teams That Work: Creating Conditions for Effective Virtual Teams*, San Francisco, CA: Jossey-Bass/Wiley, 2003.

Graetz, K., *Information Sharing in Face-to-Face, Teleconferencing, and Electronic Chat Groups*, Armstrong Lab, Wright-Patterson Air Force Base, Logistics Research Division, 1997.

Graetz, K. A., E. S. Boyle, and C. E. Kimble, "Information sharing in face-to-face, teleconferencing, and electronic chat groups," *Small Group Research*, Vol. 29, No. 6, December 1998, pp. 714–743.

Grohowski, R., C. McGoff, D. Vogel, W. B. Martz, Jr., and J. F. Nunamaker, Jr., "Implementing electronic meeting systems at IBM: Lessons learned and success factors," *Management Information Systems Quarterly*, Vol. 14, No. 4, 1990, pp. 368–383.

Hackman, J. R., and C. G. Morris, "Group tasks, group interaction process, and group performance effectiveness: A review and proposed integration," in L. Berkowitz (ed.), *Advances in Experimental Social Psychology*, Vol. 8, New York: Academic Press, 1975, pp. 45–99.

Harmon, J., "Electronic meetings and established decision groups: audioconferencing effects on performance and structural stability," *Organizational Behavior and Human Decision Process*, Vol. 61, No. 2, 1995, pp. 138–147.

Harmon, J., "Electronic meetings and intense group conflict: Effects of a policy-modeling performance support system and an audio communication support system on satisfaction and agreement," *Group Decision and Negotiation*, Vol. 7, 1998, pp. 131–153.

Harrison, D. A., S. Mohammed, J. E. McGrath, A. T. Florey, and S. W. Vanderstoep, "Time matters in team performance: Effects of member familiarity, entrainment, and task discontinuity on speed and quality," *Personnel Psychology*, Vol. 56, No. 3, September 1, 2003, pp. 633–669.

Hedlund, J., D. Ilgen, and J. R. Hollenbeck, "Decision accuracy in computer-mediated versus face-to-face decision making teams," *Organizational Behavior and Human Decision Processes*, Vol. 76, No. 1, October 1998, pp. 30–47.

Heilbronn, M., and W. L. Libby, "Comparative effects of technological and social immediacy upon performance and perceptions during a two-person game," paper presented at the meeting of the American Psychological Association, Montreal, September 1973, as cited in Williams, 1975b.

Herbsleb, J. D., and R. E. Grinter, "Splitting the organization and integrating the code: Conway's Law revisited," in *Proceedings, International Conference on Software Engineering*, Los Angeles, CA, May 16–22, 1999, pp. 85–95.

Hightower, R., and L. Sayeed, "The impact of computer-mediated communication systems on biased group discussion," *Computers in Human Behavior*, Vol. 11, No. 1, Spring 1995, pp. 33–44.

Hightower, R., and L. Sayeed, "Effects of communication mode and prediscussion information distribution characteristics on information exchange in groups," *Information Systems Research*, December 1996, Vol. 7, Issue 4, pp. 451–464.

Hiltz, S. R., *Communications and Group Decision Making: Experimental Evidence on the Potential Impact of Computer Conferencing*, Newark, NJ: New Jersey Institute of Technology Computerized Conferencing and Communications Center, Research Report 2, 1975, as cited in Whittaker, 2003.

Hiltz, S. R., and K. Johnson, "Measuring acceptance of computer-mediated communication systems," *Journal of the American Society for Information Science and Technology*, Vol. 40, No. 6, 1989, pp. 386–397.

Hiltz, S. R., K. Johnson, and M. Turoff, "Experiments in group decision-making: Communication process and outcome in face-to-face versus computerized conferences," *Human Communication Research*, Vol. 13, 1986, pp. 225–252, as cited in France, Anderson, and Gardner, 2001.

Hiltz, S. R., and M. Turoff, M, *The Network Nation: Human Communication via Computer*, Reading,MA: Addison-Wesley, 1978, as cited in Whittaker, 2003.

Hiltz, S. R., M. Turoff, and K. Johnson, "Experiments in group decision making, 3: Disinhibition, deindividuation, and group process in pen name and real name computer conferences," *Decision Support Systems*, Vol. 5, 1989, pp. 217–232.

Hinds, P. J., "The cognitive and interpersonal costs of video," *Media Psychology*, Vol. 1, 1999, pp. 283–311.

Hollingshead, A. B., "Communication and retrieval coordination in transactive memory systems," *Journal of Experimental Social Psychology*, Vol. 34, No. 5, 1998, pp. 423–442.

Hollingshead, A. B., "Information suppression and status persistence in group decision making: The effects of communication media," *Human Communication Research*, Vol. 23, No. 2, 1996a, pp. 193–219.

Hollingshead, A. B., "The rank-order effect in group decision making," *Organizational Behavior and Human Decision Processes*, Vol. 68, No. 3, December 1996b.

Hollingshead, A. B., and J. E. McGrath, "Computer-assisted groups: A critical review of the empirical research," in Richard A. Guzzo, Eduardo Salas, and Associates (eds.), *Team Effectiveness and Decision Making in Organizations*, San Francisco, CA: Jossey-Bass/Pfeiffer, 1995.

Hollingshead, A. B., J. E. McGrath, and K. M. O'Connor, "Group task performance and communication technology: A longitudinal study of computer-mediated versus face-to-face work groups," *Small Group Research*, Vol. 24, No. 3, August 1993.

Horn, D., "Is seeing believing? Detecting deception in technology mediated communication," *Proceedings of Computer-Human Interface, 2001*, Vol. 2 (extended abstracts), 2001.

Huff, C., and R. King, "An experiment in electronic collaboration," in J. D. Goodchild (chair), *Interacting by Computer: Effects on Small Group Style and Structure*, symposium conducted at the meeting of the American Psychological Association, Atlanta, GA, August 1988, as cited in Kiesler and Sproull, 1992.

Isaacs, E., and J. Tang, "What video can and can't do for collaboration: A case study," *Multimedia Systems*, Vol. 2, 1994, pp. 63–73.

Janis, I., *Victims of Groupthink*, Boston, MA: Houghton-Mifflin, 1972.

Jarvenpaa, S. L., K. Knoll, and D. E. Leidner, "Is anybody out there?: The implications of trust in global virtual teams," *Journal of Management Information Systems*, Vol. 14, 1998, pp. 29–64, as cited in Walther et al., 2001.

Jarvenpaa, S. L., and D. E. Leidner, "Communication and trust in global virtual teams," *Organization Science*, Vol. 10, 1999, pp. 791–815, as cited in Walther et al., 2001.

Jarvenpaa, S. L., and T. R. Shaw, Global virtual teams: Integrating models of trust," *Organisational Virtualness*, proceedings of the VoNet workshop, April 27–28, 1998, pp. 53–61,

Jessup, L. M., T. Connolly, and A. Tansik, "Toward a theory of automated group work: The deindividuating effects of anonymity," *Small Group Research*, Vol. 21, No. 30, 1990, pp. 333–348.

Johnson, D. L., "The effect of task difficulty on accounting-based decisions in face-to-face versus computer-mediated group settings: An experimental investigation," *Dissertation Abstracts International, Section A: Humanities and Social Sciences*, Vol. 58, No. 4-A, October 1997, p. 1357.

Joinson, A. N., " Self-disclosure in computer-mediated communication: The role of self-awareness and visual anonymity, *European Journal of Social Psychology*, Vol. 31, No. 2, March–April 2001, pp. 177–192.

Joinson, A. N., and P. R. Harris, "Causes and implication of disinhibited behaviour on the net," in J. Gackenbach (ed.), *Psychology and the Internet: Intrapersonal, Interpersonal, and Transpersonal Implications,* San Diego, CA: Academic Press, 1998.

Kelley, H., and J. Michaela, "Attribution theory and research," *Annual Review of Psychology,* Vol. 31, 1980, pp. 457–501.

Kelly, J., and T. Halvorson, "Transcripts highlight managers' flawed assumption that foam could not down shuttle," *Florida Today*, July 21, 2003.

Kiesler, S., J. Siegel, and T. McGuire, "Social psychological aspects of computer-mediated communication," *American Psychologist*, Vol. 39, 1984, pp. 1123–1134, as cited in Spears, Lea, and Lee, 1990.

Kiesler, S., and L. S. Sproull, "Response effects in electronic survey," *Public Opinion Quarterly,* Vol. 40, 1986, pp. 402–413.

Kiesler, S., and L. Sproull, "Group decision making and communication technology," *Organizational Behavior and Human Decision Processes*, Vol. 52, 1992, pp. 96–123.

Kimura, Y., and T. Tsuzuki, "Group decision making and communication mode: An experimental social psychological examination of the differences between the computer-mediated communication and the face-to-face communication," *Japanese Journal of Experimental Social Psychology*, Vol. 38, No. 2, 1998.

Kock, N., "The psychobiological model: Towards a new theory of computer-mediated communication based on Darwinian evolution," *Organizational Science*, Vol. 15, No. 3, 2004, pp. 327–348.

Krauss, R., personal communication from an unpublished study in 1976, as cited in Williams, 1977.

Krauss, R. M., and P. D. Bricker, "Effects of transmission delay and access delay on the efficiency of verbal communication," *Journal of the Acoustic Society of America*, Vol. 41, 1967, pp. 286–292, as cited in Isaacs and Tang, 1994.

Kraut, R., R. Fish, B. Root, and B. Chalfonte, "Informal communiction in organizations," in R. Baecker (ed.), *Groupware and Computer Supported Cooperative Work,* San Mateo, CA: Morgan Kaufman, 1993, pp. 287–314.

Kreumpel, K., "Making the right (interactive) moves for knowledge-producing tasks in computer-mediated groups," *IEEE Transactions on Professional Communication*, June 2000, Vol. 43, No. 2, pp. 185–195.

LaPlante, D., "Communication, friendliness, trust and the Prisoner's Dilemma," unpublished master's thesis, Windsor, Canada: University of Windsor, 1971, as cited in Williams, 1977.

Larson, J. R., Jr., C. Christensen, T. M. Franz, and A. S. Abbott, "Diagnosing groups: The pooling, management, and impact of shared and unshared case information in team-based medical decision making," *Journal of Personality and Social Psychology*, Vol. 75, 1998, pp. 93–108.

Lea, M., and R. Spears, "Computer-mediated communication, de-individuation and group decision-making," *International Journal of Man-Machine Studies*, Vol. 34, No. 2, special issue on computer-supported cooperative work and groupware: I, February 1991, pp. 283–301.

Lea, M., R. Spears, and D. DeGroot, "Knowing me, knowing you: Effects of visual anonymity on stereotyping and attraction in computer-mediated groups," *Personality and Social Psychology Bulletin*, Vol. 27, 2001, pp. 526–537.

Lim, L. H., and I. Benbasat, "The debiasing role of group support systems: An experimental investigation of the representativeness bias," *International Journal of Human-Computer Studies*, Vol. 47, No. 3, September 1997.

Manning, T. R., "Signal delay effects on rapport telepsychiatry," *Dissertation Abstracts International, Section A: Humanities and Social Sciences*, Vol. 60, No. 6-A, December 1999, p. 19–22.

Manning, T. R., E. T. Goetz, and R. L. Street, "Signal delay effects on rapport on telepsychiatry," CyberPsychology and Behavior, Vol. 3, No. 2, 2000, pp. 119–227.

Mark, G., "Conventions and commitments in distributed groups," in S. Kiesler and P. Hinds (eds.), *Distributed Work: New Research on Working Across Distance Using Technology*, Cambridge, MA: MIT Press (in press), as cited in Walther et al., 2001.

Martin, J., *Future Developments in Telecommunications*, New York: Telcom Library, 1977, as cited in Egido, 1990.

Martz, W. B., Jr., D. R. Vogel, and J. F. Nunamaker, "Electronic meeting systems: Results from field," *Decision Support Systems*, Vol. 8, 1992, pp. 141–158.

Matarazzo, G., and A. Sellen, "The value of video in work at a distance: Addition or distraction?" *Behaviour and Information Technology*, Vol. 19, No. 5, September–October 2000, pp. 339–348.

Mayer, R. C., J. H. Davis, and F. D. Schoorman, "An integrative model of organizational trust," *Academy of Management Review*, Vol. 20, No. 3, 1995, pp. 709–734.

Maznevski, M. L., and K. M. Chudoba, "Bridging space over time: Global virtual team dynamics and effectiveness," *Organization Science*, Vol. 11, No. 5, September–October 2000, pp. 473–492.

McGrath, J. E., *Groups: Interaction and Performance*, Englewood Cliffs, NJ: Prentice Hall, 1984.

McGrath, J. E., H. Arrow, D. H. Gruenfeld, A. B. Hollingshead, and K. M. O'Connor, "Group task and technology: The effects of experience and change," *Small Group Research*, Vol. 24, 1993, pp. 406–420.

McGrath, J. E., and A. B. Hollingshead, "Interaction and performance in computer assisted work groups," *Conference on Team Decision Making in Organizations*, College Park, MD: University of Maryland, January 1991, as cited in Whitworth, 1998.

McGrath, J. E., and A. B. Hollingshead, *Groups Interacting with Technology*, Thousand Oaks, CA: Sage Publications, 1994.

McGuire, T. W., S. Kiesler, and J. Siegel, "Group and computer-mediated discussion effects in risk decision making," *Journal of Personality and Social Psychology*, Vol. 52, No. 5, 1987, pp. 917–930.

McLeod, P. L. "An assessment of the experimental literature on electronic support of a group work: Results of a meta-analysis," *Human-Computer Interaction*, Vol. 7, 1992, pp. 257–280.

Mehrabian, A., *Silent Messages*, Belmont, CA: Wadsworth, 1971, as cited in Purdy and Nye, 2000.

Mehrabian, A., *Nonverbal Communications*, Chicago, IL: Aldine-Atherton, 1972, as cited in Williams, 1975b.

Meyers, D., and H. Lamm, "The group polarization phenomenon," *Psychological Bulletin*, Vol. 83, 1976, pp. 602–627.

Milgram, S., "Some conditions of obedience and disobedience to authority," *Human Relations*, Vol. 18, 1965, pp. 57–76.

Miranda, S. M., and C. Saunders, "Group support systems: An organization development intervention to combat groupthink," *Public Administration Quarterly*, Vol. 19, No. 2, 1995, pp. 193–216.

Moore, D., T. Kurtzberg, L. Thompson, and M. W. Morris, "Long and short routes to success in electronically-mediated negotiations: Group affiliations and good vibrations," *Organization Behavior and Human Decision Processes*, Vol. 77, 1999, pp. 22–43.

Morley, I. E., and G. M. Stephenson, "Interpersonal and interparty exchange: A laboratory simulation of an industrial negotiation at the plant level," *British Journal of Psychology*, Vol. 60, 1969, pp. 543–545.

Morley, I. E., and G. M. Stephenson, "Formality in experimental negotiations: A validation study," *British Journal of Psychology*, Vol. 61, 1970, p. 383, as cited in Williams, 1977.

Morris, M. W., J. Nadler, T. Kurtzberg, and L. Thompson, "Schmooze or lose: Social friction and lubrication in e-mail negotiations," *Group Dynamics*, Vol. 6, Issue 1, 2002.

Newlands, A., A. H. Anderson, J. Mullin, and A. Fleming, "Processes of collaboration and communication in desktop videoconferencing: Do they differ from face-to-face interactions?" *Proceedings of Gotalog 2000, Fourth Workshop on the Semantics and Pragmatics of Dialogue,* Goteborg, Sweden, June 2000.

Nohria, N., and R. G. Eccles, "Face-to-face: Making network organizations work," in N. Nohria and R. G. Eccles (eds.), *Network and Organizations*, Boston, MA: Harvard Business School Press, 1992, pp. 288–308.

Nunamaker, Jay F., Jr., "Future research in group support systems: Needs, some questions and possible directions," *International Journal of Human-Computer Studies*, Vol. 47, 1997, pp. 357–385.

Nunamaker, J. F., Jr., R. O. Briggs, N. C. Romano, Jr., and D. Mittleman, "The virtual office work-space: Group systems web and case studies," Chapter 7-D in D. Coleman (ed.), *Groupware: Collaborative Strategies for Corporate LANs and Intranets*, Englewood Cliffs, NJ: Prentice Hall, 1997, as cited in Nunamaker et al., 1997.

Nunamaker, J., R. Briggs, D. Mittleman, D. Vogel, and P. Balthazar, "Lessons from a dozen years of group support systems research: A discussion of lab and field findings." *Journal of Management Information Systems*, Vol. 13, No. 3, Winter 1997, pp. 163–207.

Nunamaker, J. F., Jr., D. R. Vogel, A. Heminger, W. B. Martz, Jr., R. Grohowski, and C. McGoff, "Experiences at IBM with group support systems: A field study," *Decision Support Systems: The International Journal*, Vol. 5, 1989, pp. 183–196.

Ochsman, R. B., and A. Chapanis, "The effects of 10 communication modes on the behavior of teams during co-operative problem-solving," *International Journal of Man-Machine Studies*, Vol. 6, 1974, pp. 579–619, as cited in Isaacs and Tang, 1994.

O'Connaill, B., S. Whittaker, and S. Wilbur, "Conversations over video conferences: An evaluation of the spoken aspects of video-mediated communication," *Human-Computer Interaction*, Vol. 8, 1993, pp. 389–428.

O'Hara-Devereaux, M., and R. Johansen, *Global Work: Bridging Distance, Culture, and Time*, San Francisco, CA: Jossey-Bass, 1994.

Olaniran, B. A., "Group performance in computer-mediated and face-to-face communication media," *Management Communication Quarterly*, Vol. 7, 1994, pp. 256–281, as cited in Whittaker, 2003.

Olson, G., and J. Olson, "Distance matters," *Human-Computer Interaction*, Vol. 15, 2000, pp. 139–178.

Palme, J., *Electronic Mail*, Norwood, MA: Artech House Publishers, 1995, pp. 25–30.

Pettigrew, T. F., "The ultimate attribution error: Extending Allport's cognitive analysis of prejudice," *Personality and Social Psychology Bulletin*, Vol. 5, 1979, pp. 461–476.

Pinsonneault, A., and H. Barki, "Electronic brainstorming: The illusion of productivity," *Information System Research*, Vol. 10, No. 2, 1999, pp. 110–133.

Pinsonneault, A., and K. L. Kraemer, "The impact of technological support on groups: An assessment of the empirical research," *Decision Support Systems*, Vol. 5, 1989, pp. 197–216.

Poole, M. S., M. Holmes, and G. DeSanctis. "Conflict Management in a computer-supported meeting environment," *Management Science*, Vol. 37, 1991, pp. 926–953.

Post, B., "Building the business case for group support technology," *Proceedings of the 25th Annual Hawaii International Conference on Systems Science,* Maui, HI, 1992, as cited in Nunamaker et al., 1997.

Postmes, T., and R. Spears, "Deindividuation and antinormative behavior: A meta-analysis," *Psychological Bulletin*, Vol. 123, No. 3, 1998, pp. 238–259.

Postmes, T., R. Spears, and S. Cihangir, "Quality of decision making and group norms," *Journal of Personality and Social Psychology,* Vol. 80, No. 6, 2001, pp. 918–930.

Postmes, T., R. Spears, and M. Lea, "Breaching or building social boundaries? SIDE-effects of computer-mediated communication," *Communication Research*, Vol. 25, No. 6, 1998, pp. 689–715.

Postmes, T., R. Spears, and M. Lea, "The formation of group norms in computer-mediated communication," *Human Communication Research*, Vol. 26, 2000, pp. 341–371.

Postmes, T., R. Spears, and M. Lea, "Intergroup differentiation in computer-mediated communication: Effects of depersonalization," *Group Dynamics*, Vol. 6, No. 1, March 2002, pp. 3–16.

Postmes, T., R. Spears, and M. Lea, "The effects of anonymity in intergroup discussion: Bipolarization in computer-mediated groups," *Group Dynamics* (in press).

Postmes, T., R. Spears, K. Sakhel, and D. DeGroot, "Social influence in computer-mediated groups: The effects of anonymity on social behavior," *Personality and Social Psychology Bulletin,* Vol. 27, 2001, pp. 1243–1254, as cited in Spears et al., 2002.

Purdy, J. M., and P. Nye, "The impact of communication media on negotiation outcomes," *The International Journal of Conflict Management*, Vol. 11, No. 2, 2000, pp. 162–187.

Quanquan, Z., and L. Hong, "Comparison between face-to-face and computer-mediated groups on decision-making in idea-generation task," *Acta Psychologica Sinica,* Vol. 35, No. 4, 2003, pp. 492–498.

Ramash, V., and A. R. Dennis, "The object-oriented team: Coordination and communication in global virtual software development teams," *Proceedings of the 35th Hawaii International Conference on System Sciences,* Maui, HI, 2002.

Rawlins, C., "The impact of teleconferencing on the leadership of small decision-making groups," *Journal of Organizational Behavior Management,* Vol. 10, 1989, pp. 37–52, as cited in McLeod, 1992.

Reicher, S. D., "Social influences in the crowd: Attitudinal and behavioral effects of deindividuation in conditions of high and low group salience," *British Journal of Social Psychology,* Vol. 23, 1984, pp. 341–350.

Reicher, S. D., R. Spears, and T. Postmes, "A social identity model of deindividuation phenomena," in W. Stroebe and M. Hewstone (eds.), *European Review of Social Psychology,* Vol. 6, Chichester, UK: Wiley, 1995.

Reid, A.A.L., "Comparing telephone with face-to-face contact," in I. Poole (ed.), *The Social Impact of the Telephone,* Cambridge, MA: MIT Press, 1977.

Reid, A.A.L., "Electronic person-person communications," British Post Office Communications Studies Group, P/70244/RD, 1970, as cited in Reid, 1977.

Reid, F.J.M., L. J. Ball, A. M. Morley, and J.S.B.T. Evans, "Styles of group discussion in computer-mediated decision making," *British Journal of Social Psychology,* Vol. 36, 1997, pp. 241–262, as cited in Walther et al., 2001.

Reid, F.J.M., V. Malinek, and C. Stott, "The messaging threshold in computer-mediated communication," *Ergonomics,* Vol. 39, No. 8, August 1996, pp. 1017–1037.

Reid, F.J.M., V. Malinek, C. Stott, and J.S.B.T. Evans, "The messaging threshold in computer-mediated communication," *Ergonomics,* Vol 39, 1996, pp. 1017–1037, as cited in Walther et al., 2001.

Rheingold, H., *Smart Mobs: The Next Social Revolution*, New York: Perseus Publishing, 2002.

Rice, R., "Mediated group communication," in R. Rice and Associates (eds.), *The New Media,* New York: Academic Press, 1984, pp. 129–153.

Robinson, R., and R. West, "A comparison of computer and questionnaire methods of history-taking in genito-urinary clinic," *Psychology and Health,* Vol. 6, 1992, pp. 77–84.

Rocheleau, B., "Email: Does it need to be managed? Can it be managed?" paper presented at the 2001 Conference of the American Society for Public Administration, Newark, NJ, March 10, 2001.

Rogelberg, S. G., M. S. O'Connor, and M. Sederburg, "Using the step-ladder technique to facilitate the performance of audioconferencing," *Journal of Applied Psychology,* Vol. 87, No. 5, October 2002, pp. 994–1000.

Ross, L., "The intuitive psychologist and his shortcomings: Distortions in the attribution process," in L. Berkowitz (ed.), *Advances in Experimental Social Psychology*, Vol. 10, Orlando, FL: Academic Press, 1977, pp. 173–220, as cited in Thompson and Nadler, 2002.

Ryan, M. D., and J. G. Craig, "Intergroup communication: The influence of communications medium and role induced status level on mood, and attitudes towards the medium and discussion," paper presented at the meeting of the International Communications Association, Chicago, IL, 1975, as cited in Williams, 1975b.

Salancik, G. R., "Commitment and the control of organizational behavior and belief," in B. M. Staw and G. R. Salancik (eds.), *New Directions in Organizational Behavior,* Chicago, IL: St. Clair, 1977, pp. 1–54, as cited in Sia et al., 2002.

Sellen, A. J., "Remote conversations: The effects of mediating talk with technology," *Human Computer Interaction,* Vol. 10, 1995, pp. 401–444.

Shah, S., "Computer-mediated communication and integrative bargaining: The effects of visual access and technology," working paper, Pittsburgh, PA: Graduate School of Industrial Administration, Carnegie Mellon University, 1990, as cited in Kiesler and Sproull, 1992.

Shamo, G. W., and L. M. Meador, "The effect of visual distraction upon recall and attitude change," *Journal of Communication*, Vol. 24, 1969, pp. 236–239, as cited in Short et al., 1976, p. 105.

Shapiro, Norman, and Robert H. Anderson, *Toward an Ethics and Etiquette for Electronic Mail,* Santa Monica, CA: RAND Corporation, R-3283-NSF/RC, 1985.

Short, J. A., *Bargaining and Negotiation—An Exploratory Study,* Cambridge, UK: Long-Range Intelligence Division, Technical Report E/71065/SH, 1971, as cited in Williams, 1977.

Short, J. A., "Medium of communication, opinion change, and the solution of a problem of priorities" Communication Studies Group, unpublished paper, E/72245/SH, 1972.

Short, J., E. Williams, and B. Christie, *The Social Psychology of Telecommunications*, London: John Wiley & Sons, 1976.

Sia, C. L., B.C.Y. Tan, and K. K. Wei, "Group polarization and computer-mediated communication: Effects of communication cues, social presence, and anonymity," *Information Systems Research*, Vol. 13, No. 1, March 2002, pp. 70–90.

Siau, K. L., "Group creativity and technology," *Journal of Creative Behavior*, Vol. 29, No. 3, 1995, pp. 201–216.

Siegel, J., V. Dubrovsky, S. Kiesler, and T. McGuire, "Group processes in computer-mediated communication," *Organizational Behaviour and Human Decision Processes*, Vol. 37, 1986, pp. 157–187, as cited in Spears et al., 1990.

Silver, S. D., B. P. Cohen, and J. H. Crutchfield, "Status differentiation and information exchange in face-to-face and computer-mediated idea generation," *Social Psychology Quarterly*, Vol. 57, No. 2, June 1994, pp. 108–123.

Slevin, D. P., L. W. Boone, E. M. Russo, and R. S. Allen, "CONFIDE: A collective decision-making procedure using confidence estimates of individual judgments," *Group Decision and Negotiation*, Vol. 7, No. 2, March 1998, pp. 179–194.

Spears, R., and M. Lea, "Panacea or panopticon? The hidden power in computer-mediated communication," *Communication Research*, Vol. 21, 1994, pp. 427–453.

Spears, R., M. Lea, R. A. Corneliussen, T. Postmes, and W. T. Haar, "Computer-mediated communication as a channel for social resistance: The strategic side of SIDE," *Small Group Research*, Vol. 33, No. 5, October 2002, pp. 555–574.

Spears, R., M. Lea, and S. Lee, "De-individuation and group polarization in computer-mediated communication," *British Journal of Social Psychology*, Vol. 29, 1990, pp. 121–134.

Spears, R., T. Postmes, A. Wolbert, M. Lea, and P. Rogers, "*The Social Psychological Influence of ICTs on Society and Their Policy Implications*, The Hague, Netherlands: Dutch Ministry of Education, Culture and Science, report prepared for Infodrome, 2000, as cited in Spears et al., 2002.

Sproull, L., "The nature of managerial attention," in P. Larkey and L. Sproull (sds.), *Advances in Information Processing in Organizations*, Greeenwich, CT: JAI Press, 1983.

Sproull, L. S., "Using electronic mail for data collection in organizational research," *The Academy of Management Journal*, Vol. 29, 1986, pp. 159–169.

Sproull, L., and S. Kiesler, "Reducing social context cues: Electronic mail in organizational communication," *Management Science*, Vol. 32, 1986, pp. 1492–1512.

Sproull, L., and S. Kiesler, *Connections: New ways of working in the networked environment*, Cambridge, MA: MIT Press, 1991.

Stasser, G., and D. Stewart, "Discovery of hidden profiles by decision-making groups: Solving a problem versus making a judgment," *Journal of Personality and Social Psychology*, Vol. 63, 1992, pp. 426–434, as cited in Postmes et al., 2001.

Stenmark, D., "Group cohesiveness and extrinsic motivation in virtual groups: Lessons from an action case study of electronic brainstorming," *Proceedings of the 35th Hawaii International Conference on System Sciences*, Maui, HI, 2002.

Stephenson, G. M., K. Ayling, and D. R. Rutter, "The role of visual communication in social exchange," *British Journal of Social and Clinical Psychology*, Vol. 15, 1976, pp. 113–120, as cited in Straus, 1997.

Stoner, J.A.F., "A comparison of individual and group decisions involving risk, " unpublished master's thesis, Massachusetts Institute of Technology, 1961.

Storck, J., and L. Sproull, "Reducing social context cues: Electronic mail in organizational communication," *Management Science*, Vol. 332, 1986, pp. 1492–1512.

Storck, J., and L. Sproull, "Through a glass darkly: What do people learn in videoconferences?" *Human Communication Research*, Vol. 22, 1995, pp. 197–219.

Straus, S. G., "Getting a clue: The effects of communication media and information distribution on participation and performance in computer-mediated and face-to-face groups," *Small Group Research*, Vol. 27, No. 1, 1996, pp. 115–142, as cited in Cornelius, 2003.

Straus, S. G., "Technology, group process, and group outcomes: Testing the connections in computer-mediated and face-to-face groups," *Human-Computer Interaction*, Vol. 12, No. 3, 1997, pp. 227–266.

Straus, S. G., and J. E. McGrath, "Does the medium matter? The interaction of task type and technology on group performance and member reactions," *Journal of Applied Psychology*, Vol. 79, No. 1, February 1994, pp. 87–97.

Straus, S. G., J. A. Miles, and L. L. Levesque, "The effects of videoconference, telephone, and face-to-face media on interviewer and applicant judgments in employment interviews," *Journal of Management*, Vol. 27, No. 3, 2001, pp. 363–381.

Strickland, L. H., P. D. Guild, J. C. Barefoot, and S. A. Paterson, "Teleconferencing and leadership emergence," *Human Relations*, Vol. 31, No. 7, 1978, pp. 583–596.

Tang, J. C., and E. A. Isaacs, "Why do users like video? Studies of multimedia-supported collaboration," *Computer Supported Cooperative Work: An International Journal*, Vol. 1, Issue 3, 1993, pp. 163–196.

't Hart, P. T., *Groupthink in Government: A Study of Small Groups and Policy Failures*, Baltimore, MD: Johns Hopkins University Press, 1994.

Thompson, L., and J. Nadler, "Negotiating via information technology: Theory and application," *Journal of Social Issues*, Vol. 58, No. 1, Spring 2002.

Thomson, R., and T. Murachver, "Predicting Gender from Electronic Discourse," *British Journal of Social Psychology*, Vol. 40, 2001, pp. 193–208.

Turner, J. C., M. A. Hogg, P. J. Oakes, S. D. Reicher, and M. S. Wetherell, *Rediscovering the Social Group: A Self-Categorization Theory*, Oxford, UK: Blackwell, 1987, as cited in Spears et al., 1990.

Valacich, J. S., A. R. Dennis, and T. Connolly, "Idea generation in computer-based groups: A new ending to an old story," *Organizational Behavior and Human Decision Processes*, Vol. 57, 1994, pp. 448–467.

Valacich, J. S., D. Paranka, J. F. George, and J. F. Nunamaker, Jr., "Communication currency and the new media: A new dimension for media richness," Communication Research, Vol. 20, No. 2, 2002.

Vroman, K., and J. Kovacich, "Computer-mediated interdisciplinary teams: Theory and reality," *Journal of Interprofessional Care*, Vol. 16, No. 2, 2002.

Vroom, V., and P. W. Yetton, *Leadership and Decision-Making*, Pittsburgh, PA: University of Pittsburgh Press, 1973, as cited in France, Anderson, and Gardner, 2001.

Walther, J., "A longitudinal experiment on relational tone in computer mediated and face to face interaction," in J. Nunamaker and R. H. Sprague (eds.), *Proceedings of the Hawaii International Conference on System Sciences, 1992*, Vol. 4, Los Alamitos, CA: IEEE Press, 1992a, pp. 220–231, as cited in Whittaker, 2003.

Walther, J., "Time effects in computer mediated groups," in P. Hinds and S. Kiesler (eds.), *Distributed Work,* Cambridge, MA: MIT Press, 1992b, pp. 235–258, as cited in Whittaker, 2003.

Walther, J., "Anticipated ongoing interaction versus channel effects on relational communication in computer mediated interaction," *Human Communication Research,* Vol. 20, 1994, pp. 473–501.

Walther, J. B., "Computer-mediated communication: Impersonal, interpersonal, and hyperpersonal interaction," *Communication Research,* Vol. 23, 1996, pp. 3–43.

Walther, J. B., "Group and interpersonal effects in international computer-mediated collaboration," *Human Communication Research*, Vol. 23, 1997, pp. 342–369.

Walther, J. B., J. F. Anderson, and D. Park, "Interpersonal effects in computer-mediated interaction: A meta-analysis of social and anti-social communication," *Communication Research*, Vol. 21, 1994, pp. 460–487.

Walther, J. B., M. Boos, C. L. Prell, K. D'Addario, and U. Bunz, "Misattribution and attributional redirection to facilitate effective virtual groups," paper presented at the 2nd Annual Conference of the Association of Internet Researchers, Minneapolis, MN, October 2001.

Walther, J. B., and J. K. Burgoon, "Relational communication in computer-mediated interaction," *Human Communication Research*, Vol. 19, 1992, pp. 50–88.

Walther, J. B., C. Slovacek, and L. C. Tidwell, "Is a picture worth a thousand words? Photographic images in long term and short term virtual teams," *Communication Research*, Vol. 23, 2001, pp. 105–134, as cited in Walther et al., 2001.

Warkentin, M., and P. M. Beranek, "Training to improve virtual team communication," *Information Systems Journal,* Vol. 9, No. 4, 1999, pp. 271–289.

Watson, R. T., G. DeSanctis, and M. S. Poole, "Using a GDSS to facilitate group consensus: Some intended and unintended consequences," *MIS Quarterly*, Vol. 12, No. 3, September 1988, pp. 463–478.

Watt, S. E., M. Lea, and R. Spears, "How social is Internet communication? Anonymity effects in computer-mediated groups," in S. Woolgar (ed.), *Virtual Society? Get Real: The Social Science of Electronic Technologies,* Oxford, UK: Oxford University Press, 2002.

Weick, K. E., "Cosmos vs. chaos: Sense and nonsense in electronic contexts," *Organizational Dynamics,* Fall 1985, pp. 51–64.

Weisband, S., "Group discussion and first advocacy effects in computer-mediated and face-to-face decision making groups," *Organizational Behavior and Human Decision Process,* Vol. 53, 1992, pp. 352–380.

Wetherell, M. S., "Social identity and group polarization," in J. C. Turner, M. A. Hogg, P. J. Oakes, S. D. Reicher, and M. S. Wetherell, *Rediscovering the Social Group: A Self-Categorization Theory,* 1987, Oxford/New York: Blackwell, pp. 142–170, as cited in Spears et al., 2002.

Whittaker, S., "Theories and methods in mediated communication," in A. C. Graesser, S. R. Goldman, and M. A. Gernsbaeher (eds.), *Handbook of Discourse Processes,* Mahwah, NJ: Lawrence Erlbaum Associates, 2003.

Whittaker, S., D. Frohlich, and O. Daly-Jones, "Informal workplace communication: What is it like and how might we support it?" in *Proceedings of the Conference on Computer Human Interaction,* Boston: ACM Press, 1994, pp. 131–137.

Whitworth, B., *Generating Group Agreement in Cooperative Computer-Mediated Groups: Towards an Integrative Model of Group Interaction*, Ph.D. thesis, Hamilton, New Zealand: University of Waikato, 1998, with follow-up personal correspondence via e-mail.

Wichman, H., "Effects of isolation and communication on cooperation in a two-person game," *Journal of Personality and Social Psychology*, Vol. 16, 1970, pp. 114–120, as cited in Williams, 1977.

Williams, E., "Coalition formation over telecommunications media," *European Journal of Social Psychology*, 1975a, as cited in Reid, 1977.

Williams, E., "Medium or message: Communications medium as a determinant of interpersonal evaluation," *Sociometry*, Vol. 38, 1975b, pp. 119–130, as cited in Reid, 1977.

Williams, E., "Experimental comparisons of face-to-face and mediated communications: A review," *Psychological Bulletin*, Vol. 84, No. 5, 1977, pp. 973–976 (reports an unpublished study by R. Krauss).

Wilson, E. V., "Perceived effectiveness of interpersonal persuasion strategies in computer-mediated communication," *Computers in Human Behavior*, Vol. 19, No. 5, September 2003, pp. 537–552.

Witmer, D., "Practicing safe computing: Why people engage in risky computer-mediated communication," in F. Sudweeks, M. McLaughlin, and S. Rafaeli (eds.), *Networks and Netplay: Virtual Groups on the Internet,* Cambridge, MA: AAAI/MIT Press, 1998.

Woodward, B., *Bush at War*, New York: Simon & Schuster, 2002.

Yang, K., K. Hwang, P. B. Pedersen, and I. Daibo "Effects of communication medium and goal setting of group brainstorming," *Progress in Asian Social Psychology: Conceptual and Empirical Contributions*, Westport, CT: Praeger Publishers/Greenwood Publishing Group, 2003, pp. 199–215.

Yoo, Y., and M. Alavi, "Media and group cohesion: Relative influences on social pretense, task participation, and group consensus," *MIS Quarterly*, Vol. 25, No. 3, September 2001, pp. 371–390.

Young, I., *Telecommunicated Interviews: An Exploratory Study*, Communications Studies Group, E/74165/YN, 1974a, as cited in Reid, 1977.

Young, I., "Understanding the other person in mediated interactions," British Post Office Communications Studies Group, E/74266/YN, 1974b, as cited in Reid, 1977.

Young, I., "A Three-Party Mixed-Media Business Game: A Programme Report on Results to Date," British Post Office Communications Studies Group, E/75189/YN, 1975, as cited in Reid, 1977.

Zheng, J., E. S. Veinott, N. Bos, J. S. Olson, and G. M. Olson, "Trust without touch: Jumpstarting long-distance trust with initial social activities," *Proceedings of the CHI 2002 Conference on Human Factors in Computing Systems*, Minneapolis, MN, April 20–25, 2002, pp. 141–146.

Zielinski, D., "Face Value," *Presentations*, Vol. 14, No. 6, June 2000, pp. 58–64.

Zigurs, I., M. S. Poole, and G. DeSanctis, "A study of influence in computer-mediated group decision making," *MIS Quarterly*, Vol. 12, Issue 4, 1988, pp. 625–644, as cited in Brashers, Adkins, and Meyers, 1994.

Zornoza, A., F. Prieto, C. Marti, and J. M. Peiro, "Group productivity and telematic communication," *European Work and Organizational Psychologist,* Vol. 3, 1993, pp. 117–127, as cited in Zornoza, 2002.

Zornoza, A., P. Ripoll, and J. M. Peiro, "Conflict management in groups that work in two different communication contexts: Face-to-face and computer-mediated communication," *Small Group Research,* Vol. 33, No. 5, October 2002, pp. 481–508.

Zuber, J. A., H. W. Crott, and J. Werner, "Choice shift and group polarization: An analysis of the status of arguments and social decision schemes," *Journal of Personality and Social Psychology*, Vol. 62, No. 1, 1992, pp. 50–61.